BATTLESHIPS & BATTLECRUISERS
of the Royal Navy since 1861

Cdr B. R. Coward RN

BATTLESHIPS & BATTLECRUISERS

of the Royal Navy since 1861

LONDON

IAN ALLAN LTD

First published 1986

ISBN 0 7110 1573 2

Published by Ian Allan Ltd, Shepperton, Surrey; and printed by Ian Allan Printing Ltd at their works at Coombelands in Runnymede, England

Contents

Introduction 7
The Beginning 9
I Early Developments 1861-81 10
II Fleet Modernisation 1882-91 38
III *Royal Sovereign* to *Dreadnought* 1892-1905 47
IV *Dreadnought* 1906 64
V The First Battlecruisers 1908 67
VI *Dreadnought* to *Queen Elizabeth* 1909-14 69
VII Battleships for Two World Wars 1915-17 83
VIII Post-World War 1 Construction 1920-60 94
Appendices
1 Facts and Figures 102
2 Main Armament 106
3 Aircraft 113
4 Principal Actions and Losses 116
Bibliography 119
Index of Ships' Names 120

Acknowledgements
The author wishes to thank Mr David Male and the staff of the College Library, Royal Naval College, Greenwich, and the staff of the National Maritime Museum Library for their ready assistance, the staff of the Draught Room, National Maritime Museum for the provision of ship plans, and the staff of the Fleet Air Arm Museum, Yeovilton for their help with research into aircraft carried by battleships.

The assistance of the following in the provision of photographs is also gratefully acknowledged:
Department of Photographs, Imperial War Museum, London (IWM)
Maritime Photo Library, Cromer (MPL)
Real Photographs
Fleet Air Arm Museum, Yeovilton
Messrs Wright & Logan, Portsmouth
Vickers Shipbuilding & Engineering Ltd, Barrow
Vickers plc, London (Vickers)

Prints of photographs credited to the Imperial War Museum are for sale on application to its Department of Photographs, Lambeth Road, London SE1 6HZ.

Abbreviations

AA	Anti-Aircraft
BL	Breech Loading
BLR	Breech Loading Rifled
BCS	Battle Cruiser Squadron
BS	Battle Squadron
CS	Cruiser Squadron
HA	High Angle
HACP	High Angle Control Position
ML	Muzzle Loading
MLR	Muzzle Loading Rifled
nm	Nautical Miles
UP	Unrotated Projectile

Below:
Vanguard **en route to South Africa with the royal family embarked in 1947.** *IWM*

Introduction

In the Royal Navy the battleship lasted just 100 years. Throughout that period designers strove to build ships which could inflict maximum damage on the enemy whilst at the same time resisting the enemy's attack themselves. It was a continual race between the *armament* designers and manufacturers who developed more powerful guns and more effective shells, and the *armour* designers and manufacturers who attempted to provide protection against such guns. In later years the threat to the battleship came not only from enemy gunfire but also submarine torpedoes, and eventually from aircraft torpedoes, bombs and missiles also.

The term battleship has been used throughout this book to describe ships which were designed to meet the criteria of inflicting maximum damage on the enemy whilst resisting attack themselves, although the early ships were known generically as ironclads and the earliest were actually rated as frigates. This book attempts to describe the significant details of all the battleship and battlecruiser classes built for the Royal Navy from *Warrior*, completed in 1861, to *Vanguard* which was finally towed away to be scrapped in 1960. The appendices bring together information of a more general nature.

Below:
Duke of York **in action, firing her 14in guns.**
Imperial War Museum

Comparison of Sizes

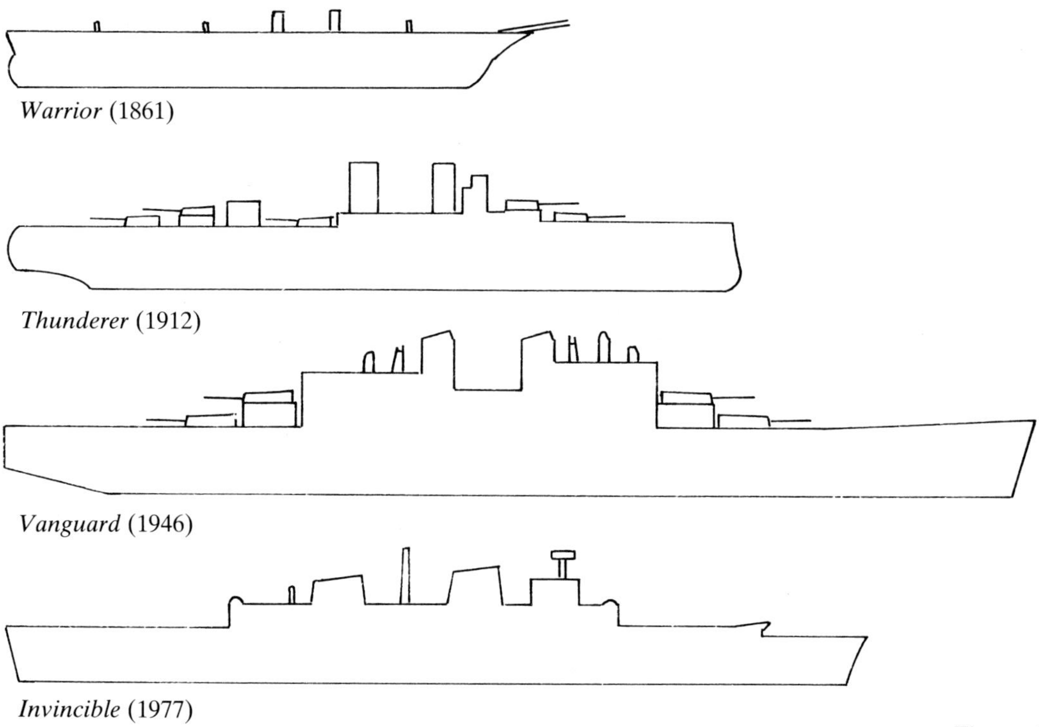

Figure 1

Class Designations

The class descriptions contain the significant details of the class members in the following format:

Displacement: imperial tons/metric tonnes
Length: feet/metres
Beam: feet/metres
Draught: feet/metres
Armour: Many different thicknesses of armour were used depending upon the part of the ship to be protected. The dimensions quoted are representative of the greatest thickness of armour fitted, given in both inches and centimetres.
Armament: The armament at build is given. Where significant changes were made during the life of a ship, these details are also indicated. In general, guns below about 6pdr are not indicated until such guns were being fitted for the purpose of close-range anti-aircraft protection. Saluting guns have been omitted. Guns' sizes have been quoted in the dimensions by which they were commonly known, eg 16in, 40mm, 12pdr, etc.
Machinery:
Fuel:
Range/Radius:
Speed: The speed given is that generally recognised to be the average maximum speed of the class.

Complement: Complements varied considerably during a ship's life and also depending upon its role. The figures quoted are representative only.

Design Notes
The significant features of the design and some of the reasoning behind it are summarised.

Historic Notes
A short summary of the career of each ship of the class is given.

Modifications
In later classes where important modifications were made, mainly in the 1930s, additional details have been included.

Note
Classes have been presented chronologically, the date of each class being taken as the date of completion of the first of class. Ships within a class are also listed in the order in which they were completed.

The Beginning

In 1850 the ships-of-the-line looked little different from those ships which had fought at Trafalgar nearly half a centruy before. They were wooden three-deckers fitted with a full rig of sails and carrying a broadside of muzzle-loading cannons. However, in the 1850s a number of developments were under way which were to render such ships out-of-date within the space of 10 years.

Steam propulsion had been in use for a number of years in smaller ships and had been to war in the Crimea where some of the block ships were steam driven. In 1851 the first British ship-of-the-line, the *Sans Pareil*, had been fitted with steam propulsion and by 1858 32 ships had been converted or built with such propulsion, although they still retained their masts and a full outfit of sails for cruising.

At this time the French were still seen as a threat to the British Isles and the newest ships of the Royal Navy were assigned to the Channel Fleet. The French had matched the developments taking place in the British Fleet and had converted a similar number of ships to steam. The French were also experimenting with other aspects of ship design, in particular in the fields of gunnery and armour. Up to the 1850s ships' guns had developed little from the smooth cannon familiar in Drake's time: they had become larger and the flintlock had been introduced as the method of igniting the powder. Breech loading had been tried and was generally unsuccessful, but the introduction of rifling in the mid-1800s was a major step forward. Rifling involved cutting grooves in a spiral inside the barrel which caused the shell to spin as it was fired. This made the shell more stable, and thus more accurate, in flight, and also allowed the shells to be shaped with pointed noses, thus increasing their penetrating power. The French experiments convinced them that the old wooden hulls were no match for such guns fired at ranges around 400-600m, the typical fighting ranges of that time, but they did show that iron plating of about 4in (10cm) thickness on the outside of the hull was sufficient to stop the shells, and from these experiments the concept of the 'ironclad' was born.

In 1858 the French ordered the first of their ironclads, *La Gloire*. There was little in her overall design which was new except for the iron plating. She was essentially a steam frigate of 5,617 tons (5,707 tonnes) displacement and 252ft (76.9m) length, carrying a full rig of sails, and 34 guns on a single gun deck which ran most of the length of the ship. This gun deck was protected by iron plating along the ship's side varying between 4.3in and 4.7in (10.9cm and 11.9cm) in thickness outside a 24in (61cm) thick wooden hull. This first order was quickly followed up by an order for three more French ironclads, *Invincible*, *Normandie* and *Couronne*.

The developments taking place on the other side of the Channel became known in London in mid-1858 and caused a certain amount of alarm. Although some of the claims about the new French ships were exaggerated, far-sighted men quickly realised that the advent of the ironclads made the British Fleet vulnerable and that wooden-hulled ships were quickly becoming obsolescent. Nevertheless there was great resistance to making changes: what was being demanded was virtually a rebuilding of the Royal Navy, a formidable undertaking both in terms of time and expense. In 1858 a number of experiments were conducted in Britain which reached much the same conclusions as the French — an iron-clad hull could withstand the guns of the day. Ideas for ironclads had been put forward as early as 1856 but had not been accepted at the time. By 1859 the Admiralty and the Government had been convinced and the order was placed for the first of the British ironclads, HMS *Warrior*.

I
Early Developments 1861-81

The period 1861-81 was one of experimentation and also attempting to keep up with new developments. Designers were trying to balance a number of factors:

1. Heavier armour and heavier guns, but smaller displacement overall in order to provide a smaller target for enemy ships.
2. High speed, which requires a long, narrow hull, coupled with manoeuvrability, which requires a shorter, broader hull.
3. A low hull with guns mounted low down for stability and a smaller target area, coupled with guns being mounted high enough for effective fire, particularly in rough weather.
4. Clear arcs of fire for the guns but a full rig of masts, sails and standing rigging for propulsion whilst cruising.

No single ship could meet all these requirements and thus every new design was something of a compromise. Different designs placed greater emphasis on different requirements so that a wide variety of ships was built with no attempt at standardisation.

Armament

In the early 1860s two main types of armament were in service. Firstly there were muzzle-loading smooth bore cannon, little different from those which had fought at Trafalgar: the 68pdr was the standard gun of this type. Also in service were breech-loading guns, also with a smooth bore, of which the main ones were the 100pdr (6in) and the 110pdr (7in). These guns were made by Armstrong but were not a success because the problems of manufacturing a satisfactory gas-tight breech mechanism had not been overcome.

Thus the Royal Navy reverted to muzzle-loading guns and also adopted rifling. The muzzle-loading rifled (MLR) gun became the standard until 1886, but these guns were massive, having very thick barrels. The problem was that gunpowder-type propellants were still used in them, and unlike cordite which burns very rapidly, gunpowder explodes and thus the guns had to be very thick to withstand the force of the explosion.

As guns developed, so did armour, and the problem was a continuing one of trying to make guns which would penetrate the ever-increasing thicknesses of armour being built into ships. The technology of armour-piercing shells was not yet developed so the only real way to improve on gun performance was to increase the size of the gun, thus allowing larger, heavier shells to be fired. Thus we see a steady increase in the calibre of guns, starting in 1866 with 9in and building up to the massive 16in guns fitted in *Inflexible* in 1881.

The handling and safety of guns was considerably improved with introduction of slide carriages. Instead of the whole gun and carriage recoiling on small wheels, the carriage was mounted on slides which were themselves fixed to the deck. The slides could rotate on rails or on a turntable to allow the gun to be trained. Improved recoil arrangements and a more precise means of elevating and training the guns became available with the introduction of Capt Scott's gun carriage and slide which entered service in 1871.

Notwithstanding the improvements in mounting arrangements of the early MLRs, the shells had to be positioned in front of the muzzle on a hoist and then rammed by hand. It was not until *Thunderer* in 1877 that the first hydraulic loading arrangements were seen at sea.

Despite being rifled, the MLRs were not very accurate. The shells had studs on them which fitted into the rifling grooves by which means the spin was imparted to the shell. Nevertheless the shells tended to be inaccurate in flight, and fighting ranges were accepted as being about 1,500yd. Guns were still trained and elevated individually, there being no central control of the armament.

Broadsides, Turrets and Barbettes

The broadside arrangement of the main armament, ie in a row along the side of the ship, had been the standard for many centuries. It was suitable for the guns and ships of the time, and also for the ranges at which battles were fought. However it did mean that the guns had to be aimed by pointing the ship, although some adjustment was possible by training guns individually. With the advent of steam propulsion in fighting ships, an enemy ship could manoeuvre out of the field of fire and, if able to 'cross the T' by crossing the bow or stern, could gain the advantage. Those who advocated the introduction of the turret, first seen in the American Civil War, suggested that it was a more flexible arrangement and, in particular,

Top:
Twin 16in turret fitted to HMS *Inflexible* (1881). *IWM*

Above:
An 11in 25-ton gun mounted in a barbette on board HMS *Temeraire* (1877). This mounting was unusual in that the gun swung back and down into the barbette, out of sight, when not firing. *IWM*

allowed the main armament to be trained on any bearing.

Monarch and *Captain* were the first ironclads with turrets, but their design was limited by a continuing insistence by the Admiralty that such ships be fully rigged. It was only when *Devastation* entered service that the advantage of the turret ship when not encumbered by masts and sails became clear to all. Nevertheless it was not until 1881 that both the broadside and the full rig became a thing of the past.

Barbettes were a variation on the turret idea. In very simple terms, a barbette is a turret without a roof. This allowed the gun itself to be mounted higher, which was considered an advantage, whilst the crew worked behind the protection of the barbette walls. Such an arrangement was reasonably successful when gunnery was conducted at short range and the trajectory of a shell was almost horizontal, but barbettes offered no protection against plunging fire.

Armour

Armour was designed to stop shells penetrating the sides of a ship. When *Warrior* was designed, trials had shown that 4½in (11.3cm) of iron plate, backed by 18in (45.7cm) of teak, was proof against the guns of the day. This armour was fitted along the sides of the ship, and was a satisfactory solution in 1861.

The design of guns soon began to outpace the armour fitted and newer ships were having to have thicker and thicker iron plate. This led to problems of weight. It became difficult to protect the whole ship completely and therefore varying thicknesses of armour were fitted, the thickest protecting the vital parts of the ship, but with thinner protection — or perhaps no protection at all — in some areas. There were obvious advantages if the vital parts of the ship could be made smaller and this was aided by concentrating all the guns in the centre of the ship in a central battery rather than having them spread out along the length of the ship. This reduced the number of guns which could be mounted but did allow those guns to be adequately protected.

The development of armour culminated in ships like *Inflexible* which had a central citadel upon which the turrets were mounted. This area of the ship was armoured with up to 24in (61cm) of iron and up to 25in (63.5cm) of teak in various combinations. The two ends of the ship were comparatively weakly protected — indeed the citadel was designed to float even if the bow and stern sections were badly damaged.

Early armour just protected a ship's side, but during this period deck armour was developed as an additional means of protecting the vitals of the ship.

Torpedoes

Robert Whitehead first produced a working torpedo in 1866. In 1871 manufacture commenced in Britain, and torpedoes became part of the armament of ironclads, firstly mounted on torpedo carriages on deck but later in torpedo tubes fitted below the waterline.

Not only did torpedoes become part of the armament, they also became something against which ships had to be protected, and thus armour below the waterline had to take such weapons into account.

Propulsion

The early ironclads were fitted with a single trunk engine driving a two-bladed propeller, which, in some ships, could be hoisted above the waterline. Although in the period 1861-81 there was little increase in speed (14kt being the average maximum speed), the period did see vast improvements in the reliability and efficiency of machinery. Efficiency was improved by increasing boiler pressure (20lb/in to 60lb/in) and the development of compound engines, the first triple compound engines being fitted in *Dreadnought*. Twin screws also became the standard.

Lines of Development

A number of lines of development can be traced in this period, many of which overlapped. For example, turret ships and broadside ships were being built together for a number of years until the turret finally took over completely. The accompanying table summarises these developments and also shows the years in which various guns, types of armour and types of propulsion first went to sea. From the table it can be seen that three lines of development were being followed in 1861. There was the 'mainstream' of ironclads starting with *Warrior* and *Black Prince*. It must be remembered that these ships were very large by the standards of the time: in fact they were so long compared with other ships of the fleet that there were only two dry docks in Britain capable of accommodating them. They were also extremely costly. Therefore there was also a move towards building smaller, slower and cheaper ironclads, of which *Defence* was the first. This line of development did not last long. Thirdly it was decided to convert a number of wooden-hulled ships then under construction to ironclads, of which *Royal Oak* was the first to be completed in 1863.

The smaller, slower ironclads and the converted wooden-hulled ships were only an interim solution and development mainly followed the *Warrior* line. All the early ships had their guns disposed in a gun deck running most of the length of the ship. However, as guns and armour became heavier, the guns were concentrated into the central battery, the first ship to adopt this new arrangement being *Bellerophon* in 1866. At about the same time there was a short programme of building extra-large ironclads, which produced *Agincourt*, *Minotaur* and *Northumberland*. The other major development in this period was, of course, the introduction of the gun turret, which first went to sea in a large ship in 1869. The last non-turret ship, *Nelson*, completed in 1881, marks the end of this period.

Battleship Development 1861-81

	BROADSIDE		
	Economy		Mainstream
1861	*Defence*		*Warrior*
1862	*Resistance*		*Black Prince*
1863			
1864	*Hector*		*Achilles*
1865			
		BATTERY	EXTRA LARGE
1866		*Bellerophon*	
1867			*Agincourt*
1868	*Valiant*	*Hercules*	*Minotaur* *Northumberland*
1869			
1870		*Audacious* *Invincible* *Vanguard*	
1871		*Iron Duke* *Sultan*	
1872		*Swiftsure*	
1873		*Triumph*	
1874			
1875			
1876			
1877		*Alexandra* *Temeraire* *Shannon*	
1878		*Northampton*	
1879			
1880		*Superb*	
1881		*Nelson*	

Note:
SB=Smooth Bore
BL=Breech Loading
MLR=Muzzle Loading Rifled

Wooden-hulled conversions	Turret	Guns	Typical armour	Propulsion
		68pdr 110pdr BL/70pdr BL 7in BL 5in BL	4½in (10.9cm) iron 18in (46.0cm) teak	Single shaft Single expansion
Royal Oak				
Prince Consort		100pdr SB		
Caledonia				
Zealous *Ocean*		9in MLR 7in MLR	5in (12.7cm) iron 10in (25.4cm) teak	
Royal Alfred		8in MLR	5½in (13.0cm) iron 10in (25.4cm) teak	
		10in MLR		
	Monarch	12in MLR 6in MLR		
Repulse	*Captain*		8in (20.2cm) iron 10in (25.4cm) teak	Twin screw Single expansion
	Devastation		12in (30.5cm) iron 18in (45.7cm) teak	
	Thunderer	11in MLR		
	Dreadnought			Twin screw Triple compound
	Neptune *Inflexible*	16in MLR	20in (50.8cm) iron 21in (53.3cm) teak	

Table I

Warrior and *Black Prince*

Displacement: 9,210 (9,357)
Length: 380 (116)
Beam: 58 (17.6)
Draught: 26 (8.0)
Armour: 4½in (11.3cm) iron, 18in (45.7cm) teak
Armament as built: 26×68pdr; 10×110pdr BL; 4×70pdr BL
Armament 1867: 28×7in MLR (24 *Black Prince*); 4×8in MLR; 4×20pdr BL
Machinery: Single screw, single expansion steam
Fuel: Coal
Range: 2,100nm at 11kt
Speed: 14kt under steam, 13kt under sail
Complement: 707

Unit	Laid Down	Completed	Builder
Warrior	May 1859	Oct 1861	Mare, Blackwall
Black Prince	Oct 1859	Sep 1862	Napier, Greenock

Design Notes
Warrior was the first British ironclad to be ordered. Her overall lines resembled those of a steam frigate but she was considerably longer, her length governed by the single gun deck required to house her 40 guns. To the length of the gun deck had to be added the bow and stern profiles which needed to be of fine lines to attain the desired speed. As such, *Warrior* was on completion the longest ship that had ever been built for the Royal Navy.

Both ships had three masts and were rigged as ships. Sail was required for cruising and for use in areas where coal was not readily available, but they were designed to be fought under steam. They were still very manpower-intensive as all hoisting of sails, weighing the anchor, hoisting boats, etc was still done by hand. In addition steering was by hand which made alterations of course a slow and laborious business. As sailing ships they were fast for their time and their performance under steam was also good. In fact, for the first British ironclads, they were a remarkably successful design. The hull of *Warrior* is still afloat!

Left:
***Warrior* in Plymouth Sound about 1862.** *IWM*

Below:
***Warrior*. Note that the upper yards on each mast have been removed and a longer bowsprit fitted.**
Maritime Photo Library

Above right:
***Black Prince* with funnels lowered.** *MPL*

Historical Notes

Warrior Channel Fleet 1861-64 and 1867-72; refitted 1864-67 and 1872-75. Coastguard ship at Portland 1875-81 and thence RNR ship in the Clyde until paying off in 1884. She served as a torpedo depot ship in Portsmouth from 1904 until she was cut down and taken to Pembroke to serve as an oil pier. However, in the early 1980s she was moved to Hartlepool to be restored and it is planned she will be on display in Portsmouth from 1986.

Black Prince Channel Fleet 1862-66 and 1875-78; Clyde 1868-74; in Reserve 1878-96. Thence moved to Queenstown as a training ship until 1910 when she returned to Plymouth, to be finally sold in 1923.

HMS *Warrior*

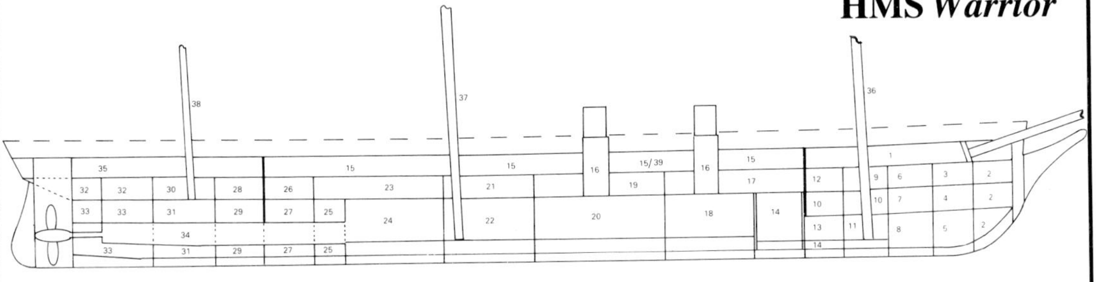

Figure 2 — HMS *Warrior* internal layout

1 Forward gun deck
2 Store
3 Gunner's room
4 Gunner's store
5 Gunner's store
6 Sick berth
7 Store
8 Sail room
9 Boatswain's, gunner's and carpenter's cabins and mess
10 Handling room
11 Boatswain's and carpenter's stores
12 Sail room
13 Cable locker
14 Magazine
15 Main gun deck
16 Funnel uptake
17 Washrooms
18 Forward boiler room
19 Bag racks
20 After boiler room
21 Sail bins
22 Shell rooms and stores
23 Chief Engineer's cabin and bag racks
24 Engine room
25 Bread room
26 Issuing and slop rooms
27 Magazine
28 Cabins
29 Provision store
30 Wardroom
31 Provisions and chest room
32 Cabins
33 Stores
34 Shaft tunnel
35 Captain's cabin and Commander's and Master's cabins
36 Foremast
37 Mainmast
38 Mizzenmast
39 Galley

Figure 2

Resistance and *Defence*

Displacement: 6,150 (6,247)
Length: 280 (76)
Beam: 54 (16.4)
Draught: 25 (7.6)
Armour: 4½in (11.3cm) iron, 18in (45.7cm) teak
Armament as built: *Resistance*: 6×7in BL; 10×68pdr; 2×32pdr
Defence: 8×7in BL; 10×68pdr; 4×5in BL
Armament 1867: 2×8in MLR; 14×7in MLR
Machinery: Single screw, single expansion steam
Fuel: Coal
Range: 1,200nm
Speed: 11kt under steam
Complement: 460

Unit	Laid Down	Completed	Builder
Resistance	Dec 1859	Jul 1862	Westwood & Baillie
Defence	Dec 1859	Dec 1861	Palmers

Design Notes
The Admiralty was concerned about the cost of ships of the size of *Warrior* and therefore ordered the building of a slower and smaller type of ironclad, but one which retained the size of gun and thickness of armour of *Warrior*. Unlike *Warrior* these ships had ram bows but were otherwise scaled down versions of the *Warrior* design. Their low speed made them unsatisfactory as they could not keep up with the larger, faster ironclads and also did not have the speed to catch enemy ships.

Above:
***Defence*. Note the ram bow and classic frigate stern.** *MPL*

Historical Notes
Defence Largely saw service around the British coast but served on the North America station 1869-70 and the Mediterranean 1871-72. Placed in Reserve 1890 and converted into a workshop in 1898.

Resistance Was the first British ironclad to serve in the Mediterranean 1864-67, otherwise served at home until paying off in 1880. In 1885 she was used as a target for gun and torpedo trials.

Royal Oak

Displacement: 6,360 (6,466)
Length: 273 (83)
Beam: 58 (17.6)
Draught: 24 (7.3)
Armour: 4½in (11.3cm) iron
Armament as built: 11×7in BLR; 24×68pdr
Armament 1867: 4×8in MLR; 20×7in MLR
Machinery: Single screw, single expansion steam
Fuel: Coal
Range: 2,000nm at low speed
Speed: 12kt
Complement: 585

Unit	Laid Down	Completed	Builder
Royal Oak	1860	Sep 1862	Chatham

Design Notes
In order to try and match the French in numbers of ironclads, it was decided to convert a number of wooden-hulled ships already under construction. *Royal Oak* was originally designed and commenced building as a two-decker but was reduced to one gun deck. She was the first ironclad to be completed in a royal dockyard. She handled well under both steam and sail.

Historical Notes
Royal Oak had only a short career, serving much of her time in the Mediterranean, and was present at the opening of the Suez Canal in 1869. She paid off in 1871 and was laid up as an economy measure, to be sold in 1885.

Below:
***Royal Oak*. The round structure between the fore and mainmasts is the funnel in its lowered position. Windsails are rigged to ventilate below decks.** *MPL*

Hector and *Valiant*

Displacement: 6,710 (6,817)
Length: 280 (76)
Beam: 56 (16.4)
Draught: 25 (8.0)
Armour: 4½in (11.3cm) iron
Armament as built: *Hector*: 4×7in BL; 20×68pdr
Valiant: 2×8in; 16×7in MLR
Armament 1867: *Hector*: 2×8in MLR; 16×7in MLR
Machinery: Single screw, single expansion steam
Fuel: Coal
Speed 12kt
Complement: 530

Unit	Laid Down	Completed	Builder
Hector	Mar 1861	Feb 1864	Napiers
Valiant	Feb 1861	Sep 1868	Westwood & Baillie and Thames Iron Works

Design Notes
These ships were ordered before *Warrior* had been to sea and when there was still considerable uncertainty about the future of ironclads. Therefore, rather than going for repeats of the *Warrior* type, in the interests of economy two smaller ships were ordered as follow-ons from *Defence* and *Resistance*. They were very similar to the 'Defences' except that they did not have a ram bow. Because of their limited size, they were really only suitable for service in home waters.

Historical Notes
Hector After one commission in the Channel Fleet (1864-67) *Hector* spent the rest of her career in the Southern Reserve until she paid off finally in 1886.

Valiant Her completion was delayed after Westwood & Baillie went broke and she was transferred to Thames Iron Works. On commissioning she went into First Reserve in Southern Ireland and remained thus until paying off in 1885. Her hull was still in use after World War 2 as an oil tank. (Note: at this time, First Reserve meant half complement.)

Right:
***Hector* in 1864 rigged as a double topsail barque.** *MPL*

Below:
***Valiant* completed as a single topsail barque, the rig to which *Hector* was also converted in 1864.** *MPL*

Achilles

Displacement: 9,820 (9,977)
Length: 380 (116)
Beam: 58 (17.6)
Draught: 27 (8.3)
Armour: 4½in (11.3cm) iron, up to 18in (45.7cm) teak
Armament as built: 14×100pdr
Armament 1865: 14×100pdr; 6×68pdr
Armament 1868: 22×7in MLR; 4×8in MLR
Armament 1874: 2×7in MLR; 14×9in MLR
Machinery: Single screw, single expansion steam
Fuel: Coal
Range: 1,800nm
Speed: 14kt
Complement: 709

Unit	Laid Down	Completed	Builder
Achilles	Aug 1861	Nov 1864	Chatham Dockyard

Design Notes
Achilles Was the first follow-on to the *Warrior* design and, as such, was the first all-iron ship to be built in a royal dockyard. She differed from *Warrior* in having a straight stem and a stern designed to provide protection for the rudder and

Above:
***Achilles* sometime after her fourth 'bow' mast had been removed and the foremast had been stepped 15ft farther forward.** *MPL*

screw. When built she was rigged with four masts, a unique arrangement, but one mast was removed early in her life. She was an exceptionally stable ship, thus making her an excellent gun platform. Overall she was a very successful ship and enjoyed a fine reputation.

Historic Notes
Achilles Spent most of her career in home waters except for three years in the Mediterranean, 1877-80. She took part in a voyage through the Dardanelles in 1878, and paid off in Plymouth in 1885.

Prince Consort, Caledonia and *Ocean*

Displacement: 6,800 (6,909)
Length: 273 (74)
Beam: 58 (17.6)
Draught: 26 (8.0)
Armour: 4½in (11.3cm) iron
Armament as built: *Prince Consort:* 7×7in BLR; 8×100pdr; 16×68pdr
Caledonia: 10×7in BLR; 8×100pdr; 12×68pdr
Ocean: 24×7in MLR
Armament 1867: All: 4×8in MLR; 20×7in MLR
Armament 1871: *Prince Consort:* 7×9in MLR; 8×8in MLR
Machinery: Single screw, single expansion steam
Fuel: Coal
Range: 2,000nm
Speed: 12kt
Complement: 605

Unit	Laid Down	Completed	Builder
Prince Consort	1860	Apr 1864	Pembroke
Caledonia	1860	Jul 1865	Woolwich
Ocean	1860	Jul 1866	Devonport

Design Notes
This class was a continuation of the policy adopted for *Royal Oak* of converting wooden hulls already under construction to ironclads. Although very similar to *Royal Oak*, these ships had more powerful, heavier machinery which, in fact, gave them no real advantage in either speed or range. They had a reputation for rolling heavily and thus were poor gun platforms. Nevertheless, they were successful ships, spending much of their time overseas.

Historical Notes
Prince Consort Channel Fleet 1864-67; Mediterranean Fleet 1867-71. Paid off and laid up at Plymouth 1871.

Caledonia Served mainly in the Mediterranean until 1872, thence in home waters until paying off 1875.

Ocean Was unique in spending the whole of her short career overseas. She served initially in the Mediterranean before going out to the China station in 1867 where she served until 1872. Thence she returned to Plymouth to pay off. She holds the record for the longest day's run by any British ironclad under sail of some 243 miles whilst en route to the Far East.

Top:
***Ocean* in 1868 rigged as a ship and with the original armament.** *MPL*

Above:
***Caledonia*.** *MPL*

Left:
***Bellerophon* with funnels lowered. The round structure just abaft the mainmast is the armoured conning tower.** *MPL*

Below:
***Zealous*.** *MPL*

Bottom:
***Royal Alfred* in Plymouth in 1867.** *MPL*

Bellerophon

Displacement: 7,550 (7,670)
Length: 300 (92)
Beam: 56 (16.4)
Draught: 26 (8.0)
Armour: 5in (12.7cm) iron, 10in (25.4cm) teak
Armament as built: 10×9in MLR; 5×7in MLR
Armament 1885: 10×8in BL; 4×6in BL; 6×4in BL; torpedo carriages
Machinery: Single screw, single expansion steam
Fuel: Coal
Speed: 14kt
Complement: 650

Unit	**Laid Down**	**Completed**	**Builder**
Bellerophon	Dec 1863	Apr 1865	Chatham

Design Notes
Bellerophon was the first ironclad to have her main armament concentrated in a battery amidships, rather than in a broadside running most of the length of the ship. This arrangement reduced the area which needed to be fully protected by armour, thus saving weight, and also made the main armament less affected by the pitching of the ship. She did not have the same fine lines as *Warrior*, being some 80ft (25m) shorter, and therefore she required more powerful engines to achieve the same speed as *Warrior*.

Historical Notes
Bellerophon served at home in the Channel Fleet 1867-71 and thereafter spent much of her career abroad. After a brief spell in the Mediterranean, she spent most of the period 1873-92 as Flagship on the North America and West Indies station. After 26 years of active service, during which she was a most happy ship, she served as a port guardship at Pembroke until 1903 when she became a stokers' training ship, being finally sold in 1922.

Zealous

Displacement: 6,100 (6,197)
Length: 252 (76)
Beam: 58 (17.6)
Draught: 25 (8.0)
Armour: 4½in (11.3cm) iron, 30in (76.2cm) teak
Armament: 20×7in MLR
Machinery: Single screw, single expansion steam
Fuel: Coal
Speed: 12kt
Complement: 510

Unit	**Laid Down**	**Completed**	**Builder**
Zealous	Oct 1859	Oct 1866	Pembroke

Design Notes
Zealous was another wooden-hulled ship converted to an ironclad. As such only minimal modifications were made from her original design and therefore she emerged as one of the shortest ironclads. She was unique at the time in only mounting one type of gun instead of having a mixed armament.

Historical Notes
Zealous served most of her career in the Pacific, based on the Canadian west coast where she spent much of her time sailing rather than steaming. She returned to home waters in 1873 and served as guardship at Southampton before finally paying off in 1875. She was sold in 1886.

Royal Alfred

Displacement: 6,700 (6,807)
Length: 273 (86)
Beam: 58 (17.6)
Draught: 27 (8.3)
Armour: 6in (15.2cm) iron
Armament: 10×9in MLR; 8×7in MLR
Machinery: Single screw, single expansion steam
Fuel: Coal
Speed: 12kt
Complement: 605

Unit	**Laid Down**	**Completed**	**Builder**
Royal Alfred	Dec 1859	Mar 1867	Portsmouth

Design Notes
Royal Alfred was yet another conversion from a wooden-hulled ship, and her completion was delayed until the earlier conversion had been tried at sea. She was similar to both *Royal Oak* and *Zealous* in layout but carried a more powerful gun armament and had thicker armour, and had her guns concentrated in a central battery, rather than a broadside as in *Royal Oak*. She handled well under sail and was as fast under sail as under steam.

Historical Notes
Royal Alfred spent much of her career on the North American and West Indies station. Her career was remarkably short, finally paying off in 1875, because of badly eroded boilers.

Minotaur and *Agincourt*

Displacement: 10,690 (10,860)
Length: 400 (122)
Beam: 59 (18.0)
Draught: 28 (8.5)
Armour: 5½in (14.0cm) iron, 10in (25.4cm) teak
Armament as built: 4×9in MLR; 24×7in MLR
Armament 1875: 17×9in MLR
Machinery: Single screw, single expansion steam
Fuel: Coal
Radius: 1,500nm
Speed: 14kt
Complement: 705

Unit	Laid Down	Completed	Builder
Minotaur	Sep 1861	Dec 1868	Thames Iron Works
Agincourt	Oct 1861	Jun 1867	Lairds

Design Notes
These ships were a direct follow-on from *Warrior* and *Achilles* but were heavier and longer. They were the first ironclads to exceed 10,000 tons displacement and, until the 'Resolution' class SSBNs were built in the 1960s, were the longest single-screw ships to serve in the Royal Navy. They retained the long gun deck of the early designs but had larger guns and thicker armour. Despite their size, no auxiliary machinery was fitted at build and they were still steered by hand! This made them slow to manoeuvre although once helm had been applied, they turned well. Power steering was fitted later. They were rigged with five masts, later reduced to three when they went into Reserve.

Historical Notes
Minotaur She served throughout her active life as Flagship of the Channel Fleet, the senior Flagship of the Royal Navy. She went into Reserve in 1887 and then, after a period as a training ship at Portland, served at Harwich as the training ship *Ganges* until finally being sold in 1922.

Agincourt In 1868 she assisted *Northumberland* with the tow of the Bermuda floating dock. Thereafter she spent the rest of her active career, like *Minotaur*, as a flagship. She was variously 2nd Flag in the Channel, then Flagship whilst *Minotaur* was refitted, and 2nd Flag in the Mediterranean. *Agincourt* paid off in 1889 and after a further 20 years as a training ship she ended her days as a coal hulk, finally going to scrap in 1960.

Below:
***Minotaur* as completed in 1868 with four square-rigged masts and the mizzen fore-and-aft rigged.** *MPL*

Bottom:
***Agincourt* with modified five-masted rig.** *MPL*

Northumberland

Displacement: 10,780 (10,905)
Length: 400 (122)
Beam: 59 (18.0)
Draught: 28 (8.5)
Armour: 5½in (13cm) iron, 10in (25.4cm) teak
Armament as built: 4×9in MLR; 22×8in MLR; 2×7in MLR
Armament 1875: 7×9in MLR; 20×8in MLR
Machinery: Single screw, single expansion steam
Fuel: Coal
Speed: 14kt
Complement: 800

Unit	Laid Down	Completed	Builder
Northumberland	Oct 1861	Oct 1868	Mare & Co

Design Notes
Northumberland was originally designed as a sister ship to *Agincourt* and *Minotaur* but her design was changed during building. Her gun deck was shorter than in the other ships — in fact, her guns were effectively in a central battery. Like the other two, she was originally rigged with five masts but this was reduced to three in 1875. She had the luxury of steam-powered steering at build.

Historical Notes
After towing the Bermuda dry dock with *Agincourt* in 1868, she served the whole of her active life in the Channel Fleet. She went into Reserve in 1890 and after a period in Reserve and thence as a training ship, was converted to a coal hulk in 1909, to be finally scrapped in 1935.

Below:
***Northumberland* as built with five masts.** *MPL*

Hercules

Displacement: 8,680 (8,820)
Length: 325 (100)
Beam: 59 (18)
Draught: 26 (7.0)
Armour: 9in (22.9cm) iron, 12in (30.5cm) teak
Armament as built: 8×10in MLR; 2×9in MLR; 4×7in MLR
Armament 1892: 6×4.7in QF added
Machinery: Single screw, single expansion steam. Converted to triple expansion in 1892
Fuel: Coal
Radius: 1,600nm
Speed: 14½kt
Complement: 630

Unit	Laid Down	Completed	Builder
Hercules	Feb 1866	Nov 1868	Chatham

Design Notes
Hercules followed *Bellerophon* in having her main armament in a central battery which was heavily armoured. The increase in armour from 4½in-5in (11.3cm-12.7cm), fitted in all the earlier ships, to 9in (22.9cm) in *Hercules*, was a major step forward, and in addition she was the first ironclad to mount the 10in gun. The gradual improvement in machinery allowed her a small increase in speed over the earlier ships. She was the earliest ironclad to undergo a major modernisation, in 1892 — her original three masts were removed, being replaced with two masts without yards or sails, and her engines were replaced with triple expansion units. However, her main armament was not improved and she was obsolescent even after 'modernisation'.

Historical Notes
Hercules had a mixed career, serving initially in the Channel Fleet 1868-74. After refit she was in the Mediterranean Fleet 1875-77 and thence in the Particular Service Squadron during the Russian Crisis of 1878. Then, after a period as a guardship and in the Reserve Fleet, she was modernised in 1892 before returning to Reserve. Like many contemporaries, she ended her days as a training ship.

Below:
***Hercules* in Plymouth in the period 1868-74. Note the indented ports which allowed two of the central battery 10in MLRs to be traversed to fire forward or aft.** *MPL*

Monarch

Displacement: 8,300 (8,433)
Length: 330 (101)
Beam: 57 (17.4)
Draught: 26 (8.0)
Armour: 10in (25.4cm) iron turrets, 6in (15.2cm) iron sides, 12in (30.5cm) teak
Armament as built: 4×12in MLR; 3×7in MLR
Armament 1871: 4×12in MLR; 2×9in MLR; 1×7in MLR
Machinery: Single screw, single expansion steam. Converted to triple expansion 1892
Fuel: Coal
Radius: 2,000nm
Speed: 14.9kt
Complement: 550

Unit	Laid Down	Completed	Builder
Monarch	Jun 1866	Jun 1869	Chatham

Design Notes
Monarch was the first large ironclad to be fitted with gun turrets, although turrets had been tried at sea on smaller coastal defence ships such as *Royal Sovereign*. However, the design was not very satisfactory as the ship carried a conventional full ship rig on three masts. This meant that the turrets had to be mounted in what was effectively a central battery and could only fire on the beam. She was the first ship to mount 12in guns. She was fitted with the most powerful engines then afloat which made her also the fastest ironclad of her day. An extensive modification was started in 1890 during which her sails were removed and her rig reduced, her machinery was replaced and a bridge fitted in front of an enlarged funnel.

Above:
***Monarch* during her second commission 1874-76.** *MPL*

Historical Notes
Monarch served in the Channel 1869-72. After refit she served again in the Channel before going to the Mediterranean in 1876. She then spent much of the next nine years in the Mediterranean, taking part in the bombardment of Alexandria in 1882, before returning once more to the Channel. After modernisation 1885-90, she then spent the rest of her days in South Africa, finally paying off in 1905 after 36 years of service.

Captain

Displacement: 7,767 (7,890)
Length: 320 (98)
Beam: 54 (16.4)
Draught: 25 (7.6)
Armour: 9in (22.8cm) iron turrets, 7in (17.8cm) iron sides
Armament: 4×12in MLR; 2×7in MLR
Machinery: Twin screw, single expansion steam
Fuel: Coal
Speed: 14kt
Complement: 500

Unit	Laid Down	Completed	Builder
Captain	Jan 1867	Jan 1870	Lairds

Design Notes
Captain was designed whilst *Monarch* was still building and was the second large ironclad to be fitted with gun turrets. Her overall layout was better as far as the guns were concerned in that the turrets were mounted one deck lower than in *Monarch*, with much of the standing rigging on the upper deck over the turrets. However, in competitive trials in 1870, *Monarch* proved to be the better ship overall. *Captain* had a very low freeboard, designed to be about 8ft (2.4m) but in practice nearer 6ft (1.8m). She was the first large ironclad to be fitted with twin screws, the first ship so fitted having been *Penelope*, an armoured corvette. The question of her stability was a subject of concern even during building.

Historical Notes
After completing in early 1870, *Captain* spent the spring and summer in home waters, during which she carried out a number of trials. In August she departed with the remainder of the Fleet for Gibraltar. On 6 September 1870 she encountered a strong gale during which, even under reefed sails, she was forced over so much that she reached her limit of stability and capsized, with the loss of all but 18 of the crew.

Above right:
***Captain* in 1870.** *IWM*

Right:
An unusual view of *Captain* in dock. Note the twin screws, the first time they had been fitted in a large ironclad. *IWM*

CAPTAIN

Repulse

Displacement: 6,190 (6,290)
Length: 252 (76)
Beam: 59 (18)
Draught: 26 (8.0)
Armour: 6in (15.2cm) iron
Armament: 12 × 8in MLR; 4 × 16in torpedo carriages
Machinery: Single screw, single expansion steam
Fuel: Coal
Speed: 12kt
Complement: 515

Unit	Laid Down	Completed	Builder
Repulse	Apr 1859	Jan 1870	Woolwich

Left:
Repulse. *MPL*

Design Notes
Repulse was the last of the wooden-hulled ships to be converted to ironclads. Although laid down at about the same time as *Royal Oak*, the first of the wooden-hulled conversions, work on *Repulse* was stopped for a number of years. By the time she was completed, her protection was far behind the standards of the day and she had no place in main fleets of the Royal Navy, hence she served her main years of active service in the Pacific. Her guns were concentrated in a central battery which left plenty of main deck space for accommodation, which was of high standard for the times. She was the last major warship built in Woolwich Dockyard.

Historical Notes
After two years in home waters as a guardship, she left for the Pacific station in 1872. She returned home in 1877 for refit after which she spent the rest of her career until 1889 as either a guardship or in Reserve. She was the only RN armoured ship ever to round the Cape Horn under sail.

'Audacious' Class

Displacement: 6,010 (6,106)
Length: 280 (86)
Beam: 54 (16.4)
Draught: 23 (7.0)
Armour: 8in (20.2cm) iron, 10in (25.4cm) teak
Armament: 10 × 9in MLR; 4 × 6in MLR
Machinery: Twin screw, single expansion steam
Fuel: Coal
Speed: 13kt
Complement: 450

Unit	Laid Down	Completed	Builder
Audacious	Jun 1867	Sep 1870	Napier
Vanguard	Oct 1867	Sep 1870	Laird
Invincible	Jun 1867	Oct 1870	Napier
Iron Duke	Aug 1868	Jan 1871	Pembroke

Design Notes
These ships were really the first ironclads to be built as a class, and were designed as smaller second-class ships for service on foreign stations. As such they were required to have a reasonable performance under sail, coal being less easily obtained abroad, and some of the distances involved being large. Accordingly they carried a very large sail area for their size. Their main armament was concentrated in a very short, two-decked central armoured battery, which placed the guns where they were least affected by the ship's motion. Consequently they were very good gun platforms.

Left:
***Iron Duke*. Note the very short central battery under the white boat amidships.** *MPL*

Historical Notes
Audacious After duties as Hull guardship 1871-74 the ship served as Flagship on the China station 1874-78. After another spell at Hull, and a refit, she returned for further service in China 1883-89. After a further refit and a third period at Hull, *Audacious* went into Reserve in 1894. She ended her career as a hulk, used variously for training or support tasks.

Vanguard Served her short career as guardship at Kingstown 1870-75. In September 1875 she was in collision with *Iron Duke* in Dublin Bay as a result of which she sank, with no loss of life.

Invincible After a brief period at Hull, *Invincible* went to the Mediterranean in 1872. Except for a refit and replacement of machinery at home, she continued to serve in the Mediterranean until 1886. Was guardship at Southampton 1886-93 after which she was used as a depot ship. She sank off Portland Bill whilst under tow in September 1914.

Iron Duke Served on the China station 1871-75. After guardship duties and a refit, she returned to the Far East in 1878 where she was Flagship until 1883. After a further refit, she completed her active service in the Channel, going into Reserve in 1890, and being finally reduced to a coal hulk in 1900.

Sultan

Displacement: 9,290 (9,440)
Length: 325 (100)
Beam: 59 (18)
Draught: 28 (8.6)
Armour: 9in (22.8cm) iron, 12in (30.5cm) teak
Armament: 8 × 10in MLR; 4 × 9in MLR; 7 × 20pdr BLR
Machinery: Single screw, two cylinder steam
Fuel: Coal
Speed: 14kt
Complement: 633

Unit	Laid Down	Completed	Builder
Sultan	Feb 1868	Oct 1871	Chatham

Above:
***Sultan* under both sail and steam in the Mediterranean.** *MPL*

Design Notes
The design was based on that of *Hercules*, having the main armament in a central battery. *Captain* and *Monarch* had not yet been tried at sea and the broadside arrangement of guns in the battery was still considered by many as the best arrangement for full-rigged ships, rather than in turrets. Although twin-screwed ships were beginning to be built, *Sultan* had only a single screw.

Historical Notes
Sultan served in the Channel 1871-76 and thence in the Mediterranean until 1879, when she went into Reserve. The ship returned to the Mediterranean in 1882 where she grounded in the Comino Channel off Malta and sank in 1889. She was eventually raised and towed back to Portsmouth. Was refitted and modernised 1893-96 although she retained her obsolescent muzzle-loading guns which made her of very little value as a fighting ship. Nevertheless she was retained in the Reserve until 1906. After various service as a training ship and repair ship, served throughout World War 2 as a depot ship for minesweepers in Portsmouth. Was finally sold after 76 years' service in 1947.

Swiftsure and *Triumph*

Displacement: 6,910 (7,010)
Length: 280 (86)
Beam: 55 (16.8)
Draught: 26 (8.0)
Armour: 8in (20.2cm) iron, 10in (25.4cm) teak
Armament: 10 × 9in MLR; 4 × 6in MLR
Machinery: Single screw, twin cylinder steam
Fuel: Coal
Complement: 450

Unit	Laid Down	Completed	Builder
Swiftsure	Aug 1868	Jun 1872	Palmers
Triumph	Aug 1868	Apr 1873	Palmers

Above:
***Swiftsure*.** *MPL*

Design Notes
These ships were very similar to the 'Audacious' class except that they were single screw and were designed with good performance under sail as a requirement. They were the last of the smaller ironclads and also the last to have guns under 10in calibre. They were fitted with hoisting propellers, a concept which was about to be phased out. The hoisting propeller was designed to reduce drag whilst under sail, and, indeed, both ships had an excellent performance under canvas.

Historical Notes
Swiftsure After service in the Channel in 1871, she went out to the Mediterranean 1872-78. Following refit at home, she was Flagship in the

Pacific 1882-85 and again 1880-90, having had another refit and a period in Reserve in between. After returning from the Pacific, was in Reserve thereafter for most of the time until 1901 when she was reduced to a stores hulk.

Triumph After early service in the Channel and Mediterranean, she went out to the Pacific 1878-82 and 1885-88. After a short spell as flagship at Queenstown, went into Reserve in 1892 until 1900. Thereafter she was used as a depot ship and a hulk in various locations, being at Invergordon from 1914. *Triumph* was finally sold in 1921.

Devastation and *Thunderer*

Displacement: 9,330 (9,480)
Length: 285 (87)
Beam: 62 (18.8)
Draught: 27 (8.3)
Armour: 14in (35.6cm) iron turrets, 12in (30.5cm) iron sides, 18in (45.7cm) teak
Armament as built: *Devastation:* 4×12in MLR
Thunderer: 2×12.5in MLR; 2×12in MLR
Armament 1891: 4×10in BL
Machinery: Twin screw, single expansion steam. Converted to triple expansion in 1891
Fuel: Coal (1,700 tons)
Radius: 4,700nm at 10kt
Speed: 13½kt
Complement: 358

Unit	Laid Down	Completed	Builder
Devastation	Nov 1869	April 1873	Portsmouth
Thunderer	Jun 1869	May 1877	Pembroke

Design Notes

These ships were a major advance in battleship design, deriving much from the early experiences with the monitors. They were the first large fighting ships not to carry a full rig of masts and sails and to depend entirely on steam for their propulsion. Considering that they were laid down only eight years after *Warrior* had first gone to sea and that broadside-armed full-rigged ships were to be built for another five years, their design was even more revolutionary.

They were regarded with a great deal of suspicion before they entered service but proved themselves to be very capable ships. Their armour broke away from the concept of protecting a central battery and returned to the fully-armoured ship, with side armour being provided for the whole length of the hull. *Thunderer* was the first ship to be fitted with a hydraulic rammer for loading the guns.

Historical Notes

Devastation In home waters 1873-74 after which she was in the Mediterranean until 1878. After a refit and a period in Reserve, was guardship at Queensferry 1885-90. After a major refit 1891-92 (during which she was re-armed and had new machinery fitted), *Devastation* spent most of the period 1893-98 either in Reserve or as a guardship in home waters. Was guardship in Gibraltar 1898-1902, after which she paid off into Reserve.

Thunderer Served in home waters 1876-78 after which she was in the Mediterranean Fleet until 1881. Went into Reserve in Malta in 1881 but was back with the Fleet 1885-86. After a period in Reserve at home, she also had a major refit for re-arming and new machinery. Returned to the Mediterranean 1891-92 but constant boiler problems reduced her to the Reserve or guardship duties at home until the end of her career in 1907.

Below left:
***Devastation* as built.** *MPL*

Below:
***Thunderer* sometime after 1890 with a fighting top added to her mast. Also, 10in breech-loading guns have replaced the original 12in MLRs.** *MPL*

Alexandra

Right:
***Alexandra* as completed.** *MPL*

Displacement: 9,490 (9,640)
Length: 325 (100)
Beam: 64 (19.6)
Draught: 26 (8.0)
Armour: 12in (30.5cm) iron, 12in (30.5cm) teak
Armament as built: 2×11in MLR; 10×10in MLR; 6×20pdr
Armament 1891: 4×9.2in BL; 8×10in MLR; 6×4in BL
Armament 1897: 4×9.2 BL; 8×10in MLR; 6×4.7in QF
Machinery: Twin screw, compound steam engines
Fuel: Coal
Speed: 15kt
Complement: 674

Unit	Laid Down	Completed	Builder
Alexandra	Mar 1873	Jan 1877	Chatham

Design Notes
Whilst designs as revolutionary as *Devastation* and *Thunderer* were under construction, traditionalists stuck to the old formula of the broadside-armed, fully rigged ship. Nevertheless *Alexandra* was a most successful design. Her main armament was carried in a central battery but this was designed so that a proportion of the guns could fire axially, ie either ahead or astern. With twin screws and more efficient machinery, her rig was unnecessary but it was not until 1889 that her masts were reduced and her sails removed. She was a handsome ship and served much of her active career as a flagship.

Historical Notes
Alexandra served continuously as Flagship, Mediterranean Fleet from 1877 to 1889. She led the Fleet into the Dardanelles in 1878 although she grounded for a period during the passage. Was also present at the bombardment of Alexandria in 1882. *Alexandra* returned home in 1889 for modernisation, during which her rig was reduced and she was re-armed. She spent the rest of her career at Portland, either as Flagship of Reserves or as guardship until 1901. Was the last broadside-armed ship to sail with the Fleet when she took part in manoeuvres in 1900.

Temeraire

Displacement: 8,540 (8,675)
Length: 285 (87)
Beam: 62 (18.7)
Draught: 27 (8.3)
Armour: 8in (20.2cm) iron barbettes, 11in (28.0cm) iron sides, 12in (30.5cm) teak
Armament: 4×11in MLR; 4×10in MLR; 2 torpedo projectors
Machinery: Twin screw, compound steam engines
Fuel: Coal
Speed: 14½kt
Complement: 550

Unit	Laid Down	Completed	Builder
Temeraire	Aug 1873	Aug 1877	Chatham

Design Notes
Temeraire was unique at the time in carrying two of her 11in guns mounted in barbettes, with the remainder in a broadside battery. The barbettes consisted of a deep circular armoured shield but without a roof, and the guns were raised by hydraulic machinery above the level of the shield for firing but lowered into the shield when stowed or when reloading. The barbette was believed to offer advantages over the turret in that the guns could be aimed more easily and the crews had more room in which to work. At the typical fighting ranges of the day (one mile or less) there was no need to shield the guns from plunging fire from above — shells travelled almost horizontally at those ranges — but as fighting ships were fitted with small, quick firing guns and small arms, so the crews in barbettes became more exposed.

In order not to obstruct the field of fire of the guns in the barbettes, she was rigged with only two masts and was thus unique in having a brig rig. However, with twin screws and more efficient machinery, she hardly needed her sails.

Historical Notes
Temeraire served virtually continuously in the Mediterranean from 1877 to 1891, except for a short period in the Channel. She was part of the fleet which made the passage of the Dardanelles in 1878 and the bombardment of Alexandria in 1882. Was in Reserve at home from 1891 to 1904 and thereafter served variously as a depot ship until being sold in 1921.

Right:
***Temeraire* under steam. Note her brig rig with only two masts and the great height of those masts.** *MPL*

Shannon

Displacement: 5,390 (5,475)
Length: 260 (79)
Beam: 54 (16.4)
Draught: 24 (7.5)
Armour: 9in (22.8cm) iron, 12in (30.5cm) teak
Armament: 2×10in MLR; 7×9in MLR
Machinery: Single screw, steam compound engine
Fuel: Coal
Speed: 12kt
Complement: 452

Unit	Laid Down	Completed	Builder
Shannon	Sep 1873	Sep 1877	Pembroke

Below:
Shannon as built. *IWM*

Design Notes
It could be argued that *Shannon* was not strictly a battleship but was, in fact, an armoured cruiser. However she was classified as a second-class battleship. She was the first ship to have a protective deck but, apart from that, her design was unsatisfactory. Her main problem was that she was too slow to work with other battleships of the time. She was intended for service on foreign stations but was not even well suited to that role.

Historical Notes
Shannon did not really enter service until 1878, her first year having been spent largely undergoing modifications. She was in the Pacific 1879-81 and thereafter spent the rest of her career either in Reserve or as a coastguard ship. Sold in 1899.

Nelson and *Northampton*

Displacement: 7,630 (7,750)
Length: 280 (86)
Beam: 60 (18.4)
Draught: 26 (8.0)
Armour: 9in (22.8cm) iron, 12in (30.5cm) teak
Armament: 4×10in MLR; 8×9in MLR
Machinery: Twin screw, steam compound engines
Fuel: Coal
Radius: 5,000nm
Speed: 13½kt
Complement: 560

Unit	Laid Down	Completed	Builder
Northampton	Oct 1874	Dec 1878	Napiers
Nelson	Nov 1874	Jul 1881	Elders

Design Notes
These two ships were the last battleships to have their guns disposed on the broadside. In some ways they were a follow-on from *Shannon* and were not a successful design. They were too small to have a proper role as Fleet battleships: their task was seen as being to engage unarmoured cruisers, but they lacked speed for this type of engagement. They had good endurance and were thus suited for foreign stations which involved long distance passages.

Historic Notes
Nelson Served on the Australia station from 1881 to 1889, part of the time as Flagship. After a long refit at home, she then served only in Reserve or as a guardship until being reduced to a hulk in 1894.

Northampton She was the Flagship on the North America and West Indies station 1879-86. Thereafter spent most of her time in Reserve at home until re-commissioned as a training ship for boys in 1894. She continued to be sea-going in this role until 1904 when she was again reduced to the Reserve and thence sold in 1905.

Below:
***Nelson* in the 1880s. Note the deep indented port beneath the foremast to allow axial fire from the main battery.** *IWM*

Bottom:
***Nelson* sometime after 1891 when her rig had been reduced to a 'military' rig. Note the quick-firing guns on the fighting tops.** *IWM*

Above:
***Dreadnought* as completed. In 1898 her funnels were raised and her bridge structure enlarged.** *IWM*

Dreadnought

Displacement: 10,886 (11,070)
Length: 320 (98)
Beam: 64 (19.6)
Draught: 27 (8.3)
Armour: 14in (35.6cm) iron turrets and sides
Armament: 4×12.5in MLR
Machinery: Twin screw, triple compound steam
Fuel: Coal
Radius: 5,700nm
Speed: 14kt
Complement: 369

Unit	Laid Down	Completed	Builder
Dreadnought	1872	Feb 1879	Pembroke

Design Notes
The name *Dreadnought* has always signified a ship of major interest in the development of warships, be it this ship, the *Dreadnought* of 1906, or the first British nuclear-powered submarine. This *Dreadnought* followed on from the design of *Devastation* and *Thunderer* but emerged as a better-balanced ship and was an example for later designs to follow. She was the first battleship to have triple compound engines and was also the last battleship to have overall armour. (The 1906 *Dreadnought* was the first battleship to have turbines.) She firmly established the steam engine as a reliable, long-range method of propulsion and the requirement for ships-of-the-line to carry sails as obsolescent.

Historic Notes
Rather strangely this fine ship started her career by going straight into Reserve. However, in 1884 she joined the fleet in the Mediterranean and remained there until 1894. After a refit, served as a coastguard ship 1895-97. After a further refit, during which she was re-boilered and had some small guns added, she again went into Reserve. From 1902 she was used mainly as a depot ship until being sold in 1908.

Superb

Displacement: 9,710 (9,865)
Length: 332 (102)
Beam: 59 (18)
Draught: 26 (8.0)
Armour: 12in (30.5cm) iron, 12in (30.5cm) teak
Armament as built: 16×10in MLR; 6×20pdr BL; 4 torpedo carriages
Armament 1891: 12×10in MLR; 10×6in BL; 4 torpedo tubes
Machinery: Single screw, steam compound engines
Fuel: Coal
Speed: 13kt
Complement: 635

Unit	Laid Down	Completed	Builder
Superb	1873	Nov 1880	Thames Iron Works

Design Notes

Superb was designed for the Turkish Government but was not handed over because of the Russo-Turkish war. She was purchased for the Royal Navy in 1880. Being built for Mediterranean and Black Sea service, she did not follow many of the British practices of design. Her guns were disposed on the broadside and she joined *Nelson* and *Northampton* as the end of the line of broadside-armed fighting ships.

Historic Notes

She served in the Mediterranean from 1880 to 1887, being present at the bombardment of Alexandria. Underwent refit and reconstruction at Chatham 1887-91 during which she was re-armed and received new engines. Thereafter she spent three years on the Clyde as a coastguard ship before going into Reserve. Was finally sold in 1906.

Below:
***Superb* in 1886.** *IWM*

Bottom:
***Superb* after reconstruction 1887-91. The modifications are typical of the period with the complete removal of yards and sails, the fitting of a 'military' rig with fighting tops, modified funnels and air intakes, new main armament and the addition of a bridge forward of the funnels.** *IWM*

Neptune

Displacement: 9,310 (9,460)
Length: 300 (92)
Beam: 63 (19.2)
Draught: 25 (7.6)
Armour: 12in (30.5cm) iron, 15in (38.1cm) teak
Armament: 4×12in MLR; 2×9in MLR; 6×20pdr BLR; 2×14in torpedo tubes
Machinery: Single screw, twin cylinder trunk engines
Fuel: Coal
Speed: 14kt
Complement: 540

Unit	Laid Down	Completed	Builder
Neptune	1874	Sep 1881	J & W Dudgeon

Design Notes
Neptune was ordered as the *Independencia* for the Brazilian Government but was taken over during the Russo-Turkish War crisis. She was a most unsatisfactory design. She had two gun turrets but the Brazilians had fitted a forecastle and a poop which meant that the turrets could only fire on the beam. She was fitted with obsolescent trunk engines at a time when triple expansion was being fitted into all other ships.

Historic Notes
She had a brief active career in the Mediterranean until 1885 when she was refitted at home. Thereafter served as a coastguard ship until 1893 when she went into Reserve. Was sold in 1903.

Above:
Neptune **in 1887 at the Spithead Review, when she had been re-masted.** *MPL*

Below:
Neptune **about 1886 when most of the yards on her mainmast had been removed.** *MPL*

Inflexible

Displacement: 11,880 (12,070)
Length: 320 (98)
Beam: 75 (22.8)
Draught: 27 (8.3)
Armour: 24in (61cm) iron citadel, 22in (56cm) iron bulkheads, 17in (43.2cm) turrets
Armament: 4×16in MLR; 6×20pdr; 2 submerged torpedo tubes
Machinery: Twin screw, steam triple compound
Fuel: Coal
Speed: 14.75kt
Complement: 440

Unit	Laid Down	Completed	Builder
Inflexible	Feb 1874	Oct 1881	Portsmouth

Design Notes

Inflexible was designed with a heavily-armoured 'impregnable' citadel which was capable of floating even if the unarmoured bow and stern sections of the ship were heavily damaged. She carried the thickest armour which has ever been fitted to a British fighting ship, and also carried the heaviest guns then in service. Her wide beam of 75ft (22cm) was not matched until the 'Royal Sovereign' class in 1892. She was the first ship to be fitted with submerged torpedo tubes, and also had electric lighting.

Her arrangement below decks was extremely complex with such ideas as tanks packed with cork being incorporated into her protection. She was one of the last ships to be rigged and to carry sails, although she had no need for them. *Inflexible* was really the last of an era and was a rather ugly ship.

Historic Notes

She served her first four years in the Mediterranean, which included taking part in the bombardment of Alexandria, during which she was hit by a shell, causing some damage. In 1885 was refitted during which her rig was reduced. Was in Reserve 1885-90, after which she returned to the Mediterranean until 1893. Thereafter she served as either a guardship or in Reserve until being sold in 1903.

Below:
***Inflexible* sometime after 1885 when her rig had been reduced. The two turrets are between the funnels.** *MPL*

II
Fleet Modernisation 1882-91

The short period from 1882 to 1891 saw a number of significant developments and a move towards a standard design for battleships which was to last for another 50 years. The ship which emerged from this period had its main armament on the centreline forward and aft, with a superstructure in between which contained a battery of secondary armament, the armoured conning tower and upon which were mounted the funnels, masts, bridge, searchlights, etc. The short period in which it was considered important that all the main guns should be able to fire directly forward or aft had passed; it was now accepted that end fire was only possible from half the main armament.

Armament

These years saw the return to breech-loading guns. The breech loaders of the early 1860s had not been successful because a satisfactory breech mechanism had not been developed, but by the 1880s these problems had been overcome and large BL guns came into service. In addition slower-burning powders were being developed to supersede explosive gunpowder. The advantage of the new powders was that it was possible to develop guns with greater muzzle velocity, and therefore greater penetrating power. Such guns required longer barrels in which to accelerate the shells, a development which in itself made muzzle-loading less and less attractive.

The threat from torpedo boats increased significantly in this era. Torpedo boats were small and fast and could carry the much-improved torpedoes then coming into service, and it was found that battleships at night, particularly at anchor, were very vulnerable to torpedo boat attack. The smaller calibre guns available in the early 1880s were of little use against this threat. However, the development of quick-firing (QF) guns during this decade made effective fire against fast-moving targets possible and, to some extent, reduced the torpedo boat threat. At the same time, battleships themselves were carrying more torpedoes, some in tubes below the waterline, some in tubes or carriages above.

Left:
The after barbette twin 12in mounting in HMS *Collingwood*. *IWM*

Turrets and Barbettes

The broadside arrangement of the main armament had gone, and the discussion over whether the turret or barbette was better continued. The major class built during this period, the 'Admirals', had barbettes but the issue was still not decided by 1891.

Armour

Armour developed little in this period, not matching gun development, although the introduction of compound armour plate which was iron plate with a steel face was a step forward. At the same time, it was impossible to completely armour the whole ship so the layout of the armour protection was a balance between protecting the important parts of the ship, and not adding too much weight. Therefore different thicknesses of armour were applied to different parts of the ship, some areas not being armoured at all. In addition it was being realised that the effectiveness of gunfire, plus speed and manoeuvrability, were in themselves factors which added to the overall protection of the ship.

Propulsion

Three-cylinder compound engines had been first introduced in *Dreadnought*. This period saw the more efficient triple expansion engines coming into service and a modest increase in speed to just over 17kt. Forced draught became available in most ships, which increased engine power.

The Navy in 1891

By 1891 the only battleships in active service which were still of the old style of design were *Inflexible* and *Temeraire*. The remainder of the battleships serving in the Channel and Mediterranean Fleets had been completed since 1881. The largest fleet of battleships in 1891 was in the Mediterranean (10 ships) with a small contingent in the Channel (four ships), one on the North America and West Indies station, and the remainder either in Reserve or as coastguard ships. Battleships in this period could spend a large proportion of their time in Reserve: indeed some ships went into Reserve immediately after completion, often being commissioned for the annual summer manoeuvres and returning to the Reserve for the winter. There were various levels of Reserve such as Fleet Reserve, Dockyard Reserve, etc, and as a ship grew older, she tended to be relegated to the lower levels of Reserve.

Tactical thought changed in this period. After some years when end-on fire by battleships was considered important and ramming was still considered a possible tactic, by the end of the 1880s it was doctrine that battleships would fight in line ahead. Fighting ranges were still short, of the order of 2,000yd, at which range both the main armament and secondary battery could be used against the target. The main armament was designed to penetrate the enemy's armour whilst the smaller guns struck at the lightly-armoured sections of the enemy ship. As time went on, the secondary armament also became the main protection against torpedo boats, with quick-firing guns important in this role.

Torpedoes also were becoming a significant part of the battleship armament, usually with the tubes arranged to fire on the beam. Protection underwater against torpedoes received some consideration and plans were drawn up for double bottoms which would allow the torpedo to explode against a thin outer shell, whilst a thicker, inner hull remained intact. However, such ideas were not to be put into practice for another 20 years or so.

The end of this decade saw a significant development in the battleship fleet — the ordering of the 'Royal Sovereign' class.

Ajax and *Agamemnon*

Displacement: 8,510 (8,655)
Length: 280 (86)
Beam: 66 (20.2)
Draught: 24 (7.3)
Armour: 18in (45.7cm) iron citadel, 16in (40.7cm) compound turrets, 18in (45.7cm) teak
Armament: 4×12.5in MLR; 2×6in BL
Machinery: Twin screw, steam triple compound
Fuel: Coal
Speed: 13kt
Complement: 345

Unit	Laid Down	Completed	Builder
Ajax	Mar 1876	Mar 1883	Pembroke
Agamemnon	May 1876	Mar 1883	Chatham

Design Notes
These two ships marked the beginning of a new era — the days of the broadside-armed ship was passing and also the last fully-rigged ship-of-the-line had been built. These ships were designed as scaled-down 'Inflexibles' but without masts and sails. The main requirement was for a smaller draught suited to operations in the Baltic or Black Sea, but the design was not a success. In particular the citadel was not capable of floating if either the bow or stern sections were badly damaged. Also they handled extremely badly, their steering being very erratic until the lines of the stern were modified. They were the last battleships to be fitted with muzzle-loading rifled (MLR) guns, and the first to have a secondary battery. In addition battleships were started to be fitted with smaller guns as protection against torpedo launches.

Historic Notes
Ajax Spent most of the period 1885-91 as coastguard ship at Greenock. The remainder of her undistinguished career was in Reserve, finally being sold in 1904.

Agamemnon Most of her active service was abroad. She was on the China station 1884-86, after which she returned to the Mediterranean. Spent most of the period 1886-92 in the Mediterranean apart from a short period of nine months detached to the East Indies station. She returned home in 1892 to go into Reserve and was sold in 1903.

Below:
***Ajax*, probably late in her career.** *IWM*

Conqueror and *Hero*

Displacement: 6,300 (6,400)
Length: 270 (82)
Beam: 58 (17.6)
Draught: 25 (7.6)
Armour: 12in (30.5cm) compound belt, 12in (30.5cm) compound turrets, 13in (33.0cm) teak
Armament: 2×12in BL; 4×6in BL; 7×6pdr QF; 6×14in torpedo tubes
Machinery: Twin screw, steam triple compound
Fuel: Coal
Speed: 14kt
Complement: 330

Unit	Laid Down	Completed	Builder
Conqueror	Apr 1879	Mar 1886	Chatham
Hero	Apr 1884	May 1888	Chatham

Design Notes
These ships were designed and built at a time when ramming was still seen as a realistic tactic, although the fact that ramming called for superior speed and good manoeuvrability was largely forgotten. The layout of these two ships only allowed them to engage targets ahead. They had little to commend them and they spent virtually no time in active service.

Historic Notes
Conqueror Joined the Reserve in 1887 and then became a gunnery training ship in 1889 based in Devonport. Apart from occasional manoeuvres, she spent the rest of her time in Devonport, finally paying off in 1902.

Hero Was also a gunnery training ship but in Portsmouth, where she remained until 1905, when she went into Reserve. She ended her days by being sunk as a target ship in 1908.

Right:
***Conqueror*. Note the very restricted field of fire of the single turret.** *IWM*

Imperieuse and *Warspite*

Displacement: 8,500 (8,635)
Length: 315 (96)
Beam: 62 (18.8)
Draught: 27 (8.2)
Armour: 10in (25.4cm) compound belt, 8in (20.2cm) compound barbettes, 10in (25.4cm) teak
Armament: 4×9.2in BL; 10×6in BL; 6 torpedo tubes
Machinery: Twin screw, steam triple compound
Fuel: Coal
Radius: 5,500nm
Speed: 16.7kt
Complement: 555

Unit	Laid Down	Completed	Builder
Imperieuse	Aug 1881	Sep 1886	Portsmouth
Warspite	Oct 1881	Jun 1888	Chatham

Design Notes
Although *Imperieuse* was completed almost a year ahead of *Collingwood*, she was laid down one year after, and therefore some of the design features of these two ships followed on from ideas already incorporated in *Collingwood*. These included the layout of the armoured protection with only a belt of armour protecting the centre of the ship, and also the mounting of the main armament in barbettes. The ships also incorporated a few ideas borrowed from the French, in particular in having the main armament in single barbettes, one forward, one aft and one on each side of the ship. The pronounced tumble-home was also a feature of French ships. *Imperieuse* left the builders with two masts and rigged as a brig, but her masts and sails were quickly removed after sea trials proved her ineffective under sail.

Historic Notes
Imperieuse Was largely in Reserve from completion until 1889 when she went out to the China station. Returned home in 1894 for another short period in Reserve and a refit, after which she went to the Pacific as Flagship 1896-99. Thereafter she served at home, variously in Reserve, as a depot ship and briefly in the Home Fleet. Sold in 1913.

Warspite Mainly in Reserve until 1890 when she went to the Pacific as Flagship. She returned home in 1893 and remained in Reserve or as a guardship until 1899 when she relieved *Imperieuse* in the Pacific. Returned home to go into Reserve in 1902 and was sold in 1905.

Top:
Imperieuse as built with a brig rig. This was quickly removed after sea trials. *IWM*

Above:
Warspite. Note the main armament in single barbettes and the pronounced tumblehome amidships. *IWM*

Colossus and *Edinburgh*

Displacement: 9,150 (9,290)
Length: 325 (100)
Beam: 68 (20.8)
Draught: 26 (98.0)
Armour: 18in (45.7cm) compound citadel, 16in (40.7cm) compound turrets, 22in (56.0cm) teak
Armament: 4×12in BL; 5×6in BL; 2×14in torpedo tubes
Machinery: Twin screw, steam compound engines
Fuel: Coal
Speed: 16kt
Complement: 396

Unit	Laid Down	Completed	Builder
Colossus	Jun 1879	Oct 1886	Portsmouth
Edinburgh	Mar 1879	Jul 1887	Pembroke

Design Notes
These ships followed the *Inflexible* design in having a massively armoured central citadel but weakly protected bow and stern sections. They were the last ships to have this short-lived layout. In other respects they were leaders of important developments, being the first ships to be built almost entirely of steel and with compound armour in general use: *Ajax* and *Agamemnon* had used some compound armour, consisting of steel-faced iron, in their turrets, but in *Colossus* and *Edinburgh* it was used for the turrets and citadel. These ships also marked the return to breech-loading guns, having the first large ones in battleships since the 'Armstrongs' of the 1860s. With their turrets mounted one on each side of the centreline, and the narrow superstructure forward and aft, it was possible for these ships to fire their guns ahead and astern, but the arrangement was far from ideal.

Historic Notes
Colossus After a period in home waters, she went out to the Mediterranean in 1887 and remained there until 1893. Was coastguard ship at Holyhead 1893-1901, after which she went into Reserve. She enjoyed a short period again as a guardship 1904-06, after which she was sold.

Edinburgh After taking part in the Jubilee Review in 1887, *Edinburgh* served with the Mediterranean Fleet from late 1887 to 1894. Then she was coastguard ship at home until 1897 after which she went into Reserve for two years. From 1899-1905 was a gunnery ship at Sheerness. In 1908 used as a target for gun trials, designed to test modern armour and armour piercing shells.

Below:
Colossus. *IWM*

Collingwood

Displacement: 9,500 (9,650)
Length: 325 (100)
Beam: 68 (20.8)
Draught: 26 (8.0)
Armour: 18in (45.7cm) compound belt, 11½in (28.2cm) compound barbettes, 20in (50.8) teak
Armament: 4×12in BL; 6×6in BL; 4 torpedo tubes
Machinery: Twin screw, triple compound steam
Fuel: Coal
Radius: 7,000nm
Speed: 16.8kt
Complement: 500

Unit	Laid Down	Completed	Builder
Collingwood	Jul 1880	Jul 1887	Pembroke

Design Notes

Collingwood can be considered to be the first in a line of development of battleships which extended until 1946. She returned to the layout which had been seen in *Dreadnought* and which was to become a standard for all subsequent classes to follow, although there were variations over the years. Her main armament was situated on the centreline forward and aft, an arrangement which finally accepted that it was not necessary for battleships to be able to bring all their main armament to bear either ahead or astern. Between the two pairs of main armament was the superstructure which contained the secondary battery of 6in guns, the air intakes to the boiler rooms, and the funnels, and upon which were mounted the bridge, the armoured conning tower, masts, searchlights, small arms, etc. Her armoured protection started the move away from the massive central citadel. The barbettes were fully protected but the centre of the ship only was armoured with a belt along the waterline, protecting the boilers and engines. The 6in battery was not armoured. An armoured deck helped protect the vitals of the ship but she was designed to be able to float and fight with compartments forward or aft flooded — coal bunkers placed close to the ship's side in the bow and stern sections afforded some protection to these areas.

Her guns were mounted in barbettes, on the grounds that barbettes were simpler and lighter than turrets, that they allowed the guns to be mounted higher, and also the guns had more freedom in depression and elevation. She was also the fastest battleship of her time and had a longer range than her contemporaries. Despite some criticism of the design at the time, she was a marked step forward in battleship development and also a handsome ship.

Historic Notes

Collingwood, despite her qualities, was placed in Reserve after completion, where she remained for most of the time up to 1889. Went to the Mediterranean Fleet from 1889 to 1897, after which she returned home for service as a coastguard ship. Again into Reserve in 1903 and was finally sold in 1911.

Below:
***Collingwood*. The lack of protection for the gun crews from not only enemy gunfire but also sea and spray is evident.**
IWM

Bottom:
***Collingwood* late in her career when torpedo net defences had been fitted. The booms can be seen along the ship's side.**
MPL

'Admiral' Class

Displacement: 10,450 (10,615)
Length: 330 (101)
Beam: 68 (20.8)
Draught: 28 (8.6)
Armour: 18in (45.7cm) compound belt, 14in (35.6cm) compound barbettes
Armament: *Benbow:* 2×16.5in BL; 10×6in; 5 torpedo tubes
Others: 4×13.5in BL; 6×6in; 5 torpedo tubes
Machinery: Twin screw, steam triple compound
Fuel: Coal
Speed: 17kt
Complement: 530

Unit	**Laid Down**	**Completed**	**Builder**
Rodney	Feb 1882	Jun 1888	Chatham
Benbow	Nov 1882	Jun 1888	Thames Iron Works
Anson	Apr 1883	May 1889	Pembroke
Camperdown	Dec 1882	Jul 1889	Portsmouth
Howe	Jun 1882	Jul 1889	Pembroke

Design Notes
The 'Admirals' were effectively repeats of the *Collingwood* design although there were variations within the class. For example, *Howe* and *Rodney* were slightly shorter than the other ships, while *Benbow* was unique in mounting the new 16.5in breech loading guns, singly in barbettes. The smaller guns of the other ships were mounted in pairs in the barbettes. The class's armour protection was similar to *Collingwood's*.

These ships were the first true battleship class — the term ironclad was no longer appropriate. They confirmed the trends started with *Collingwood* of the main armament mounted on the centreline forward and aft, no sail power, and the return to large breech-loading guns.

Historic Notes
Rodney After serving in home waters 1888-94, she was in the Mediterranean for the following three years. From 1897-1901 was a coastguard ship at home, after which she went into Reserve before being sold in 1909.

Benbow Served her first three years 1888-91 in the Mediterranean. Was then in Reserve until 1894 when she went to Greenock as the coastguard ship. Remained there until 1904 when she returned to the Reserve, finally being sold in 1909.

Anson She was Rear Admiral Flagship in the Channel 1889-93, after which served in the Mediterranean until 1900. Thereafter she was in Reserve until 1901 when re-comissioned for the Home Fleet. She paid off in 1904 to Reserve and was sold in 1909.

Camperdown Was Flagship briefly in the Mediterranean 1889-90 and then Flagship in the Channel until 1892. Then returned to the Mediterranean and in the summer of 1893 was in collision with, and sank, *Victoria*. After repair and refit, she spent the rest of her years at home, variously in Reserve or as a coastguard ship until

Below:
***Rodney*. The ports for the 6in battery can be seen along the side of the superstructure.** *MPL*

Left:
***Anson* with two fighting tops on the mainmast.** *MPL*

she went as an accommodation ship for submarines at Harwich 1908-11.

Howe Joined the Channel Fleet in 1890 and during manoeuvres in the Mediterranean in 1892 she ran aground and was only salvaged with difficulty. After repair she was back in the Mediterranean 1893-96, after which remained at home, either in Reserve or as a coastguard ship, until her sale in 1910.

Sans Pareil and *Victoria*

Displacement: 10,470 (10,635)
Length: 340 (104)
Beam: 70 (21.3)
Draught: 29 (8.8)
Armour: 18in (45.7cm) compound belt, 17in (43.2cm) turret, 7in (17.8cm) teak
Armament: 2×16.25in BL; 1×10in BL; 12×6in BL; 12×6pdr; 4 torpedo tubes plus 2 torpedo carriages
Machinery: Twin screw, triple expansion steam
Fuel: Coal
Radius: 7,000nm
Speed: 17kt
Complement: 430

Unit	Laid Down	Completed	Builder
Victoria	Apr 1885	Mar 1890	Elswick
Sans Pareil	Apr 1885	Jul 1890	Thames Iron Works

Design Notes
After the 'Admirals', these two ships must be considered a retrograde step. It is not entirely clear why they were designed as they were, with a limited main armament which could only fire ahead of the beam. For protection astern they had a single, unprotected 10in gun. The completion of both ships was delayed due to hold-ups in the production of the 16.25in guns. These ships represented a reversion to gun turrets after the barbettes of the 'Admiral' class.

Although three-cylinder compound engines had been at sea for some time, these ships were the first to be fitted with triple expansion engines. They were not considered a particularly successful pair of ships.

Historic Notes
Victoria After delays with the fitting of her guns, she went to the Mediterranean in 1890 as the Flagship. Ran aground in January 1892, after which she was refitted in Malta. Returned to the Fleet as the Flagship of Vice-Admiral Sir George Tryon. She was sunk in a collision with *Camperdown* on 22 June 1893 during Fleet manoeuvres off Tripoli.

Sans Pareil After a short period in Reserve, she served in the Mediterranean 1892-95. After this short period abroad, she spent the rest of her career at home, either in Reserve or as a guardship. Finally sold in 1907.

Above right:
***Sans Pareil*. The guns of the 6in battery can be seen over the white boat.** *MPL*

Right:
***Victoria*. The single 10in gun can be seen at the stern.** *IWM*

Nile and *Trafalgar*

Displacement: 12,590 (12,790)
Length: 345 (105)
Beam: 73 (22.2)
Draught: 29 (8.8)
Armour: 20in (50.8cm) compound belt, 18in (45.7cm) compound turrets, 10in (25.4cm) teak
Armament: 4×13.5in BL; 6×4.7in QF; 8×6pdr; 6 torpedo tubes
Machinery: Twin screw, triple expansion steam
Fuel: Coal
Radius: 6,500nm at 10kt
Speed: 17kt
Complement: 577

Unit	Laid Down	Completed	Builder
Trafalgar	Jan 1886	Mar 1890	Portsmouth
Nile	Apr 1886	Jul 1891	Pembroke

Design Notes
The design of these two ships lay somewhere between the 'Admirals' and *Dreadnought*. The layout of the armoured protection was similar to the 'Admirals' but otherwise the design was very much a repeat of *Dreadnought*. They represented a large increase in displacement for battleships, being the first to exceed 12,000 tons. They were also the first to carry quick firing (QF) guns which were a significant improvement in defence against torpedo boats.

Historic Notes
Trafalgar Between 1890 and 1897 she was the 2nd Flagship in the Mediterranean, then port guardship at Portsmouth until 1902, after which she was placed in Reserve. *Trafalgar* became a drill-ship at Sheerness in 1907 and reverted to the Reserve in 1909, being sold in 1911.

Nile Also spent most of her active career in the Mediterranean, from 1891 to 1898. She was then a guardship in Devonport until 1903, after which she remained in Reserve until the end of her career in 1911.

Below:
***Trafalgar* in 1890, with short funnels.** *MPL*

Bottom:
***Nile* with her characteristic tall funnels.** *IWM*

III
Royal Sovereign to *Dreadnought* 1892-1905

This era is commonly known as the 'pre-dreadnought' period and, indeed, the battleships built at this time became to be known as pre-dreadnoughts themselves. It was an era when Britain and the British Empire was the major world power and much of the strength of the country was seen to rest upon a powerful Royal Navy. The battleships built in this period represented a steady improvement in design with no major variations in that design overall. Many of the ships built in the late 1890s and early 1900s served successfully in World War 1, although in secondary roles: the large force of ships which took part in the Dardanelles campaign was composed almost entirely of ships built in these years.

Gunnery

The 'Royal Sovereigns' had 13.5in guns mounted in barbettes, but the standard gun of this period quickly became the 12in. Armoured gun shields were built over the barbettes, and these soon developed into all-round protection for the gun and gun crew. This arrangement remained the standard for battleships for the rest of the battleship era. In fact, the armoured gun shield has become known as a gun turret and this terminology is used for the remainder of this book. Breech-loading mechanisms were perfected, and all-round loading was developed, along with loading systems which could load the guns at any angle of elevation. Cordite, a slow-burning powder, became the standard propellant and longer guns allowed the use of higher muzzle velocities. The battleship gun had come of age.

However, whilst the guns themselves had progressed considerably, the effectiveness of gun fire was still very poor. Guns were still aimed individually, range finding was inaccurate and the gun sights themselves were of poor quality. It was in 1899 that Capt Percy Scott and others started to develop the techniques of gunnery, Scott introducing his ideas of 'continuous aim' where guns continually followed the target. Fire control was still very rudimentary, with range measured from the conning tower and passed to the guns by voice pipe; the range finders themselves were small. Nevertheless this period saw fighting ranges increasing to about 8,000yd, greater emphasis on gunnery practices with competitions being introduced, and an overall improvement in gunnery standards.

The 6in gun and the 9.2in gun were used initially as part of the main armament to engage enemy battleships. However, the increase in fighting ranges and improvements in armoured protection were to lead to the main armament becoming of just one size, 12in or larger, with the secondary armament becoming more and more important as protection against torpedo-armed vessels. Originally the torpedo boat was small and not very sea-worthy: as protection against them, the torpedo boat destroyer was developed as a fast, manoeuvrable, small ship, which itself developed to the stage where it became not only a defence against torpedo boats but itself took over the role of the torpedo boat. Thus the destroyer armed with guns and torpedo tubes could be used in both attack and defence and it was against such vessels that the secondary armament of battleships was required.

Armour

Iron armour disappeared completely in this period and was replaced by ever-improving types of steel armour. Steel was much stronger than iron and thus better protection could be provided for less weight. Nevertheless armour-piercing shells were also improving and many of the older ships were very vulnerable by the end of their active careers, the advance in guns and shells having overtaken the capability of their armoured protection.

Propulsion

The triple expansion engine reigned supreme. There were steady improvements in the power and efficiency of engines and boilers which allowed a modest increase in speed to around 19kt and improvements in range. Oil burning was introduced although only as an addition to coal: the oil was sprayed on to the coal; it could not be used as the only fuel. In addition forced draught improved the efficiency of the boilers although there were many problems with early forced-draught boilers. Such boilers could also be steamed using just natural draught, although this reduced the power developed and thus reduced the speed of the ship.

The Fleet in 1905

The emphasis of the Navy moved from the Mediterranean to home waters in the mid-1900s, because the power of Germany was increasing and the need for a stronger Navy in the Channel and North Sea became apparent. The names of the various fleets in home waters and the status within them varied in this period: ships served variously in the Home, Channel, Atlantic and Reserve Fleets, sometimes with full crews, sometimes with nucleus crews. Lessons were of course learned from events abroad, particularly the Sino-Japanese War of 1894-95, the Spanish-American War of 1898 and the Russo-Japanese War of 1904-05.

The form of the Fleet as a whole became clear in this period. The battleship carried the large guns and was well protected with armour; lighter cruisers with minimal protection were the eyes of the Fleet, seeking out the enemy and using their superior speed to stay out of trouble; and the destroyer was not only the defence against enemy destroyers but also a vessel of attack itself. From the wars mentioned above, the effectiveness (or otherwise) of gunfire, and the possibility of fighting at longer range than was previously envisaged, became apparent. The importance of the heavy guns of the battleship too was clear: the smaller guns made little impact on enemy battleships. The advantage of speed in gaining and maintaining the tactical advantage was re-emphasised, and lessons were learned in improving armoured protection. Wireless telegraphy was also used for the first time in war.

'Royal Sovereign' Class

Displacement: 14,150 (14,375)
Length: 380 (116)
Beam: 75 (22.8)
Draught: 28 (8.6)
Armour: 18in (45.7cm) compound and steel belt, 17in (43.2cm) compound and steel barbettes
Armament: 4×13.5in BL; 10×6in QF; 16×6pdr QF; 7 torpedo tubes
Machinery: Twin screw, steam triple expansion
Fuel: Coal
Radius: 4,700nm
Speed: 15½-18kt (*Royal Sovereign* only ship to reach 18kt)
Complement: 712

Unit	Laid Down	Completed	Builder
Royal Sovereign	Sep 1889	May 1892	Portsmouth
Empress of India	Jul 1889	Aug 1893	Pembroke
Ramillies	Aug 1890	Oct 1893	Thomsons
Resolution	Jun 1890	Nov 1893	Palmers
Revenge	Feb 1891	Mar 1894	Palmers
Repulse	Jan 1890	Apr 1894	Pembroke
Royal Oak	May 1890	Jun 1894	Lairds

Design Notes

This class marked a consolidation of the design lessons learned in the previous decade and also an advance. Earlier ships had suffered considerably from their low freeboards, which made guns difficult to work effectively in rough weather and also slowed the ships down when making into a head sea, because of the weight of water running across the decks. The 'Royal Sovereigns' had their freeboard increased by the addition of another deck, although the barbettes were kept at much the same height as in previous designs so that the extra deck came up almost level to the top of the barbettes. This made it look as if the guns had been mounted much lower than in previous barbette classes, whereas in fact the guns were still mounted comparatively high. The guns were largely unchanged from earlier designs: they still had to return to a fixed heading for reloading. When first built, four of the 6in guns were mounted in casemates; the remainder were also placed in casemates during reconstruction at the turn of the century. (A casemate is a compartment within which secondary armament is mounted and within which the gun can be trained and elevated.)

Steel armour was starting to come into service and this class was the first to be fitted with it, giving the ships an overall mix of compound and steel armour. Their maximum speed could only be reached with forced draught supplying air to the boilers. In practice this arrangement did not work well and they normally propelled in natural draught, which reduced them to about 15½kt.

They were very fine, handsome fighting ships which set a standard for the next decade.

Historical Notes

Royal Sovereign Channel Fleet 1892-97 (Flagship 1892-95). She went to the Mediterranean 1897-1902 and thence as a guardship until 1905. After refit in 1903, she was in Reserve until 1907 when she joined the Home Fleet. Sold in 1913.

Empress of India Channel Fleet 1893-97 after

which she went to the Mediterranean until 1901. After a short period in Ireland, she was refitted in 1902, after which she was either in Reserve or with the Home Fleet until 1911. After a further period in Reserve, she was used as a target ship, being sunk in the Channel in 1913.

Ramillies She was in the Mediterranean 1893-1903, after which was refitted at Chatham. Thereafter she was in Reserve or with the Home Fleet until 1911. Sold in 1913.

Resolution Channel Fleet 1893-1901. Then after a short period as a coastguard ship and in Reserve, she was refitted in 1902. Thereafter was a guardship or in Reserve until 1909 when she served briefly with the Home Fleet. Paid off 1911 and sold in 1914.

Revenge Had a long and distinguished career. After service at home until 1896, she was in the Mediterranean until 1900. After a short period as a guardship and in Reserve, she was refitted 1902-05. Thereafter served as a gunnery ship until 1912. Was in Reserve 1913-14 when she returned to active service and was used for bombardments off the Belgian coast and for gunnery trials. Renamed *Redoubtable* in 1915. Withdrawn from service in 1917, she was finally sold in 1919.

Repulse Served continuously in the Channel 1894-1902. After a short period in the Mediterranean she was refitted in 1903. Thereafter she remained at home, mainly in Reserve until being sold in 1911.

Royal Oak In Reserve until 1896, then went to the Mediterranean until 1902. After refit, she was with the Home or Channel Fleet until 1905 and subsequently mainly in Reserve until being sold in 1914.

Above:
***Resolution*. The higher freeboard forward and aft made ships of this class much drier than earlier classes.** *MPL*

Bottom:
***Empress of India*.** *MPL*

Hood

Displacement: 14,150 (14,375)
Length: 380 (116)
Beam: 75 (22.8)
Draught: 28 (8.6)
Armour: 18in (45.7cm) compound and steel belt, 17in (43.2cm) compound and steel turrets
Armament: 4×13.5in BL; 10×6in QF; 10×6pdr QF; 7×18in torpedo tubes
Machinery: Twin screw, steam triple expansion
Fuel: Coal
Radius: 4,720nm
Speed: 16½kt
Complement: 692

Unit	Laid Down	Completed	Builder
Hood	Aug 1889	May 1893	Chatham

Design Notes

Hood A 'Royal Sovereign' in all respects except that, in effect, her upper deck was removed so that her main armament could be mounted in turrets. She therefore suffered from the same problems as all earlier turret ships with a low freeboard making for poor sea-keeping and limiting the use of the main armament.

She was the last battleship to have her guns in turrets, as the term was originally understood. Subsequently guns mounted in barbettes were covered by an armoured gun house which, in time, came to be known as a turret. Hence there is sometimes confusion over the terminology.

Historic Notes

Hood served the first part of her career in the Mediterranean 1893-1900. After a short refit and spell as guardship at home, she returned to the Mediterranean until 1902. After a refit she was with the Home Fleet or in Reserve until 1910. Next went to Queenstown until 1911, and thereafter was used briefly as a target ship. In 1914 her hull was used as a blockship at Portland; it remains there to this day.

Below:
***Hood* in later years. Note how flimsy were the ships' bridges and wheelhouses of this era.** *IWM*

Barfleur and *Centurion*

Displacement: 10,500 (10,665)
Length: 360 (110)
Beam: 70 (21.4)
Draught: 26 (8.0)
Armour: 12in (30.5cm) compound belt, 9in (22.6cm) compound and steel barbettes, 6in (15.2cm) steel gunhouses
Armament: 4×10in BL; 10×4.7in QF; 8×6pdr; 7 torpedo tubes
Machinery: Twin screw, steam triple expansion
Fuel: Coal
Radius: 6,000nm
Speed: $18\frac{1}{2}$kt (forced draught)
Complement: 620

Unit	Laid Down	Completed	Builder
Centurion	Mar 1890	Feb 1894	Portsmouth
Barfleur	Oct 1890	Jun 1894	Chatham

Design Notes
These two ships were designed specificially as second-class battleships primarily for service on the remote foreign stations, such as China. In some ways they were scaled-down 'Royal Sovereigns' with smaller armament and reduced armour. Their secondary armament was particularly light. Both ships were considerably reconstructed just after the turn of the century when 6in guns replaced the 4.7in.

Historic Notes
Centurion Flagship on the China station 1894-1901. After a two-year reconstruction at Portsmouth, she returned to the China station again in 1903-05. Thereafter was in Reserve at home before finally paying off in 1909 and being sold in 1910.

Barfleur After a year in Reserve, she went to the Mediterranean 1895-98 and thence to China until 1902. Following reconstruction, she spent most of her time in Reserve at home before being sold in 1910.

In 1900 both ships were involved in the Boxer Rising in China and landed parties for operations against the Boxers. A certain Capt Jellicoe, later to be Admiral of the Fleet Earl Jellicoe, was the Flag Captain in *Centurion*, and the future Admiral of the Fleet Earl Beatty was Commander of *Barfleur*.

Below:
***Barfleur* after reconstruction 1902-04.** *MPL*

Renown

Displacement: 12,350 (12,545)
Length: 380 (116)
Beam: 72 (22.0)
Draught: 27 (8.3)
Armour: 8in (20.2cm) Harvey steel belt, 10in (25.4cm) Harvey steel barbettes, 6in (15.2cm) Harvey steel casemates
Armament: 4×10in BL; 10×6in QF; 12×12pdr; 5×18in torpedo tubes
Machinery: Twin screw, steam triple expansion
Fuel: Coal
Radius: 8,500nm
Speed: 18kt (19.7kt trials with forced draught)
Complement: 674

Unit	Laid Down	Completed	Builder
Renown	Feb 1893	1895	Pembroke

Design Notes
Renown was considered a second-class battleship although she was 2,000 tons heavier than *Centurion* and *Barfleur*. Athough her main armament was rather light, the secondary battery was very strong. She was the first battleship to have all-steel armour and also the first to employ an angled armoured deck. This 2in (5.1cm) armoured deck, which ran the width of the ship at about waterline level, was angled downwards at the sides to meet the armoured belt. *Renown* was fast for her day.

Historic Notes
Renown had a short but distinguished career, having a reputation as a smart ship, good at ceremonial. She was Flagship for the Jubilee Review in 1897 and then went to the North America and West Indies station as the Flagship of Admiral Sir John Fisher. She transferred to the Mediterranean in 1899 where she was again Flagship, remaining until 1902. In 1902 *Renown* took the Duke and Duchess of Connaught to India to attend the Delhi Durbar. After a refit and a period in Reserve, she repeated the voyage to India in 1905, this time with the Prince and Princess of Wales (the future George V and Queen Mary). Went into Reserve on return from India and ended her days as a stokers' training ship. She was sold in 1914.

Below:
***Renown*. Note the comparatively modern shape of the forward turret.** *IWM*

'Majestic' Class

Displacement: 14,900 (15,140)
Length: 390 (120)
Beam: 75 (22.8)
Draught: 27 (8.3)
Armour: 9in (22.8cm) Harvey steel belt, 14in (35.6cm) Harvey steel barbettes, 10in (25.4cm) Harvey steel gun shields
Armament: 4×12in; 12×6in; 16×12pdr; 5×18in torpedo tubes
Machinery: Twin screw, steam triple expansion
Fuel: Coal plus oil
Speed: 17kt
Complement: 672

Unit	Laid Down	Completed	Builder
Majestic	Feb 1894	Dec 1895	Portsmouth
Magnificent	Dec 1893	Dec 1895	Chatham
Victorious	May 1894	Nov 1896	Chatham
Prince George	Sep 1894	Nov 1896	Portsmouth
Jupiter	Apr 1894	May 1897	Clydebank
Mars	Jun 1894	Jun 1897	Lairds
Caesar	Mar 1895	Jan 1898	Portsmouth
Hannibal	May 1894	Apr 1898	Pembroke
Illustrious	Mar 1895	Apr 1898	Chatham

Design Notes
In outward appearance rather similar to the 'Royal Sovereigns', this large class (the most numerous battleship class ever built) incorporated a number of significant developments. The slight reduction in gun size to 12in was more than offset by the improvements in gun design and mounting. These guns used slow-burning cordite as the propellant, and with stronger, longer barrels, a greater muzzle velocity was possible. The barbettes had fully-

Above:
***Majestic* early in her career.** *MPL*

Below:
***Jupiter*. Compare her drab, grey appearance with that of *Majestic*.** *Real Photos*

Below:
***Prince George*. The sloped armour for the forward turret can be seen.** *MPL*

enclosed shields over them, with 10in (25.4cm) sloping armoured fronts. In fact, what is now known as a gun turret had arrived. These turrets also had improved loading arrangements, although in all ships except *Caesar* and *Illustrious* loading was still from a fixed fore-and-aft loading position. The latter two ships had the improved Whitworth BIII mounting which allowed all-round loading. The use of all-steel armour meant that better protection could be provided without a weight penalty.

Mars was the first battleship to burn oil fuel, and all but *Jupiter* and *Illustrious* were subsequently fitted for oil burning. The oil was sprayed directly on to the coal so the boilers could not be steamed on oil alone.

Altogether the 'Majestics' were fine ships and they all saw service in World War 1, although, in some cases, in supporting roles only.

Historic Notes

Majestic Flagship of the Channel Fleet 1895-1904. Thereafter served variously in the Channel, Atlantic and Reserve Fleets until being refitted in 1907-08. Thence she served mainly with the Home Fleet until 1914. Joined 7 BS August 1914 and acted as convoy escort and Channel and Dover patrol. In early 1915 joined the force off the Dardanelles. Sunk 27 May 1915 by torpedo fired from a U-boat.

Magnificent 2nd Flagship Channel Fleet 1895-1904. After a short refit in 1904 served with Channel, Atlantic and Reserve Fleets until refit 1907-08. Thence with Home Fleet until becoming a training ship in 1911. Joined 9 BS 1914. Turrets removed 1915 for use in monitors. Used as a trooper in late 1915, and thereafter in miscellaneous supporting roles only. Sold 1921.

Victorious Served in the Channel until 1898, then in the Mediterranean briefly before going out to China. Returned to the Mediterranean 1900-03 before refit at Chatham. Thereafter she served at home, variously with the Channel, Atlantic and Home Fleets. In 1914 was sent to the Humber but moved to Scapa Flow in 1916 as a base ship for dockyard workers who assisted with the maintenance of the Grand Fleet.

Prince George Channel Fleet 1896-1903. Refitted 1904, and served mainly with the Home Fleet until 1914 when she joined 7 BS for operations in the Channel. In early 1915 she joined the Dardanelles campaign and was badly damaged 2 May 1915. After repair in Malta, she returned to the Dardanelles and was hit by a torpedo which failed to explode. After the end of the Dardanelles campaign, she returned to Chatham for dockyard and depot ship duties. Sold in 1921.

Jupiter Channel Fleet 1897-1905. Thence mainly with the Home Fleet until refit in 1909. Served as a gunnery training ship in 1912 before going into Reserve. In 1914 joined 7 BS and was guardship on the Humber and Tyne. In early 1915 was sent as an icebreaker to Archangel in the White Sea to assist in keeping the port open to allow supplies to be sent to the Russian army. In the summer of 1915 she returned to the Channel and thence on for service in the Middle and Far East. She was used for special service 1917-18 and finally paid off in 1919.

Mars Channel Fleet 1897-1905. Refitted in 1905 when she fitted as the first oil-burning battleship. Thereafter she served in the Channel, Reserve and Home Fleets until 1914 when she was sent to the Humber. In late 1915 her turrets were removed for use in a monitor and she went out to the Mediterranean as a trooper. Paid off in 1916 and became a depot ship at Invergordon.

Caesar Mediterranean Fleet 1898-1903. Thereafter saw service in the Channel, Atlantic and Home Fleets until 1914. Joined 7 BS for duties in the Channel, and in 1915 went to North America and West Indies station where she remained until 1918. 1918-19 was in the Mediterranean and Black Seas. Sold 1921.

Hannibal Channel Fleet 1898-1904. Thereafter Channel, Atlantic or Reserve Fleets until 1914 when she was sent to the Humber and then Scapa Flow as a guardship. In late 1915 she went to the Mediterranean as a trooper and her turrets were transferred to monitors. Thereafter she went to the East Indies and then Middle East until the end of the war. Sold in 1920.

Illustrious Mediterranean Fleet 1898-1904 then Channel Fleet until 1908 and thence into Reserve until 1914. Initially served as a guardship in various locations, then in 1915 was either on detached service or with the Channel Fleet. Went to the Tyne as an ammunition ship 1916-17 and then as a store ship in Portsmouth until 1919. Sold in 1920.

'Canopus' Class

Displacement: 12,950 (13,155)
Length: 390 (120)
Beam: 74 (22.6)
Draught: 26 (8.0)
Armour: 6in (15.2cm) Krupp steel belt, 12in (30.5cm) Krupp steel barbettes, 8in (20.2cm) Krupp steel turrets
Armament: 4×12in; 12×6in; 10×12pdr; 4×18in torpedo tubes
Machinery: Twin screw, steam triple expansion. First water tube boilers
Fuel: Coal
Speed: 18kt
Complement: 682

Unit	Laid Down	Completed	Builder
Canopus	Jan 1897	Dec 1899	Portsmouth
Ocean	Feb 1897	Feb 1900	Devonport
Goliath	Jan 1897	Mar 1900	Chatham
Glory	Dec 1896	Oct 1900	Lairds
Albion	Dec 1896	Jun 1901	Thames Iron Works
Vengeance	Aug 1898	Apr 1902	Vickers

Design Notes
These ships carried the same main armament as the 'Majestics' but in other ways were similar to *Renown*. They were the first battleships to be fitted with Krupp steel armour which was about 50% stonger than Harvey steel and three times stronger than iron armour. This meant that the armour protection could be made thinner to achieve the same strength, and thus also became lighter overall. With very similar dimensions, these ships were 2,000 tons lighter than the 'Majestics'.

They were also the first battleships to be fitted with water tube boilers. The advantage of such boilers was that they were smaller, lighter and more efficient than the 'locomotive' boilers used previously. This class was able to make 18kt without the assistance of forced draught air supply to the new boilers.

Historic Notes
The class as a whole saw much action in the first part of World War 1 and formed a significant part of the battleship force during the Dardanelles campaign.

Canopus Mediterranean Fleet 1899-1903. Thence variously in the Atlantic, Channel, Reserve and Home Fleets until 1908 when she returned to the Mediterranean for 18 months. Thereafter was in the Home Fleet until the outbreak of war. Was detached to the South Atlantic and was off the coast of Chile during the Battle of Coronel but was unable to join the action which resulted in the sinking of the *Good Hope* and *Monmouth* by German forces. *Canopus* then returned to Port Stanley, Falkland Islands where she was the guardship during the Battle of the

Below:
***Vengeance* being launched at Barrow.**
Vickers Shipbuilding & Engineering Ltd

Falkland Islands in December 1914. Although *Canopus* engaged the German forces from inside the harbour, she was in no condition to join the subsequent successful chase. She joined the Dardanelles campaign early in 1915 where she remained until 1916. *Canopus* served as an accommodation ship at Chatham for the remainder of the war.

Ocean In 1900 served initially in the Mediterranean but then went out to China 1900-05. Channel Fleet 1906-08, thence mainly with the Home Fleet until 1914. Early in the war she went out to the East Indies for convoy duties and thence to operations in the Persian Gulf. In early 1915 she joined the Dardanelles force, and during the major attempt to force the Straits on 18 March 1915 she hit a mine and was also damaged by shell fire. In the early evening she was abandoned — she sank later that evening.

Goliath China station 1900-03. Thereafter she served with the Channel, Home and Reserve Fleets until the outbreak of war. She took part in the operations off Ostend early in World War 1 and then went briefly to the East Indies. In the spring of 1915 joined the Dardanelles force, and was sunk by torpedoes from a torpedo boat on 13 May 1915.

Glory Flagship China station 1900-05; Channel Fleet 1906. Thence Reserve or Home Fleet until joining Mediterranean Fleet 1907-09, thereafter Home Fleet until 1914. At the start of the war she served on the North America and West Indies station, until joining the Mediterranean Fleet in mid-1915. In 1916 she went to North Russia and remained there until 1919, when she went as depot ship at Rosyth. Sold 1922.

Albion Served in China 1901-05. Thereafter Channel, Home and Reserve Fleets until 1914. Early in the war she went out to Cape station but then joined the Dardanelles force in 1915. After spending 1915 in the Dardanelles, she served on the East Coast of England 1916-18, finally becoming an accommodation ship at Devonport. Sold 1919.

Vengeance In Mediterranean 1902, thence to China 1903-05. Channel Fleet 1906-08. Thereafter with the Home Fleet until 1913 when she became a gunnery training ship. In 1914 she served in the Atlantic before going to the Dardanelles early in 1915. From late 1915 to 1917 she served variously in Egypt, the East Indies and on the East African coast. Became an ordnance depot in Devonport 1918. Sold 1921.

Refits

All the ships were extensively refitted in the period 1904-12 during which engines and boilers were improved and, more importantly, a fire control system was added for gunnery. Radio communications were also fitted.

Below:
***Vengeance* after 1912 with the upper fighting top on the foremast adapted for gunnery control.** *MPL*

'Formidable' Class

Displacement: 15,000 (15,240)
Length: 400 (124)
Beam: 75 (22.9)
Draught: 27 (8.3)
Armour: 9in (22.8cm) Krupp steel belt, 10in (25.4cm) Krupp steel turrets
Armament: 4×12in; 12×6in; 16×12pdr; 4×18in torpedo tubes
Machinery: Twin screw, steam triple expansion
Fuel: Coal
Radius: 8,000nm
Speed: 18kt
Complement: 780

Unit	Laid Down	Completed	Builder
Formidable	Mar 1898	Sep 1901	Portsmouth
Irresistible	Apr 1898	Feb 1902	Chatham
Implacable	Jul 1898	Sep 1901	Devonport

Design Notes
For their day, the 'Formidables' were very large and powerful ships. They can be considered as improved 'Canopuses', displacing an extra 2,000 tons. They were armed with the new 12in Mk IX gun which was a 40 calibre gun as opposed to 35 calibre in the earlier Marks (ie the barrel was 40 times the bore of the gun in length — 40×12in). *Irresistible* was fitted with the BVIII mounting which allowed loading at all angles of elevation, as well as at all angles of training. This was effectively the final development of the twin 12in mounting — the later Marks just represented detailed improvements.

Historical Notes
Formidable Mediterranean Fleet 1901-08. After a refit in 1909, served mainly with Home Fleet until 1912 when she joined 5 BS. On 1 January 1915, early in the morning, the squadron was steaming off Start Point when *Formidable* was struck by a torpedo. A second torpedo from a submarine hit her about 45 minutes later after which she slowly sank.

Irresistible Mediterranean Fleet 1902-08. After refit, she joined the Channel Fleet in the latter part of 1908. 1909-11 Home Fleet; 1912-14 5 BS. Joined the Dardanelles force early in 1915, and during the major action on 18 March 1915 she hit a mine, was abandoned and finally sank under gunfire.

Implacable Mediterranean Fleet 1901-08. Channel 1909, thence to Atlantic Fleet 1909-12. 5 BS 1912-14. Joined Dardanelles force early 1915 but was then detached to Italian Fleet to assist with blockade of Austria May 1915. Thereafter served on the East Indies to Egypt route before returning to the Mediterranean 1916-17. Joined the Northern Patrol 1918-19. Sold 1921.

Below:
***Implacable*. Four of the 6in guns can be seen along the ship's side.** *MPL*

'London' Class

Displacement: 15,000 (15,240)
Length: 400 (124)
Beam: 75 (22.9)
Draught: 26 (8.0)
Armour: 9in (22.8cm) steel belt, 10in (25.4cm) steel turrets
Armament: 4×12in; 12×6in; 16×12pdr; 4×18in torpedo tubes
Machinery: Twin screw, steam triple expansion
Fuel: Coal
Radius: 8,000nm
Speed: 18kt
Complement: 730

Unit	Laid Down	Completed	Builder
Bulwark	March 1899	March 1902	Devonport
London	Dec 1898	June 1902	Portsmouth
Venerable	Jan 1899	Nov 1902	Chatham
Queen	March 1901	March 1904	Devonport
Prince of Wales	March 1901	March 1904	Chatham

Top:
***Bulwark*. The layout of the secondary armament is well illustrated with four 6in guns at main deck level beneath the superstructure, two more on the upper deck at either end of the superstructure, and four 12pdrs centrally placed within the superstructure on the upper deck. Gun ports for two more 12pdrs can be seen below the forecastle; the other two 12pdrs on the port side fired through ports beneath the quarterdeck.** *MPL*

Above:
***Prince of Wales* in Portsmouth.** *Real Photos*

Design Notes
These ships were very similar to the 'Formidable' class except for some changes in the armoured protection. *Queen* and *Prince of Wales* are sometimes considered as a separate class; the differences in these two ships were minor.

Historic Notes
Bulwark Flagship in Mediterranean 1902-07, thence refit at home. Mainly Home Fleet 1909-12 then joined 5 BS. Blew up at Sheerness whilst loading ammunition 26 November 1914 and was totally destroyed.

London Was Flagship of the Coronation Review 1902 then went out to the Mediterranean until 1907. Thence to refit in Chatham followed by service with Channel, Atlantic and Home Fleets. Joined 5 BS 1912. At the outbreak of war served in the Channel until early 1915 when she joined the Dardanelles force. She remained in the Mediterranean thereafter until joining the Reserve 1919. Sold in 1920.

Venerable Mediterranean Fleet 1902-09. Refit Chatham 1909 thence Atlantic Fleet until 1912 when she joined 5 BS. In late 1914 and early 1915 was used for bombardments of Belgian coast. Joined Dardanelles force May 1915 and then became part of the force blockading the Austrian Navy in the Adriatic.

Queen Mediterranean Fleet 1904-08. After refit in Devonport, was with Home Fleet until 1912; 5 BS 1912-14. Joined Dardanelles force early 1915 then joined the Italian Fleet in May 1915, remaining as a base ship in Taranto until the end of the war. Sold 1920.

Prince of Wales Mainly in Mediterranean 1904-07. Thence Atlantic Fleet and Home Fleet until 1912; 5 BS 1912-14. Accompanied Dardanelles force early 1915, then joined force blockading Austrian Fleet until 1918.

'Duncan' Class

Displacement: 14,000 (14,225)
Length: 405 (124)
Beam: 75 (22.6)
Draught: 26 (8.0)
Armour: 7in (17.8cm) belt, 10in (25.4cm) turrets
Armament: 4×12in; 12×6in; 10×12pdr; 4×18in torpedo tubes
Machinery: Twin screw, steam triple expansion
Fuel: Coal
Speed: 19kt
Complement: 720

Unit	Laid Down	Completed	Builder
Russell	Mar 1899	Feb 1903	Palmers
Exmouth	Aug 1899	Jun 1903	Lairds
Duncan	Jul 1899	Oct 1903	Thames Iron Works
Montagu	Nov 1899	Oct 1903	Devonport
Albemarle	Jan 1900	Nov 1903	Chatham
Cornwallis	Jul 1899	Feb 1904	Thames Iron Works

Design Notes
These ships were developments of the 'Formidables' but in order to obtain extra speed they were made slightly longer and also their displacement was reduced by some sacrifice in armoured protection.

Historic Notes
Russell Mediterranean Fleet 1903-04; then Home, Channel and Atlantic Fleets until 1909. She then returned to the Mediterranean until 1912, thence Home Fleet. In 1914 became Flagship of 6 BS in the Grand Fleet. Transferred to the Channel Fleet November 1914 and took part in bombardments of the Belgian coast, including the U-boat pens at Zeebrugge. Went out to the Mediterranean in November 1915 and was mined and sunk off Malta 27 April 1916.

Exmouth Mediterranean Fleet 1903-04, thence Flagship of the Home Fleet until 1907. Flagship in the Mediterranean 1908-12 when she returned to the Home Fleet. Gunnery training ship 1913. Joined 6 BS 1914 but transferred to Channel Fleet November 1914 for bombardment of the Belgian coast. In May 1915 joined the Dardanelles force and remained in the Mediterranean until 1917 when she went into Reserve. Sold 1920.

Duncan Mediterranean Fleet 1903-05 and thence Channel and Atlantic Fleets until 1908. Returned to the Mediterranean until 1912 when she joined the Home Fleet. Initially 6 BS in 1914 until transferred to the Channel Fleet in November 1914. Joined the Mediterranean Fleet in 1915 but was not involved in major Dardanelles operations. Into Reserve 1917 and sold 1920.

Montagu Mediterranean Fleet 1903-05, thence Channel Fleet. She ran aground on Lundy Island 30 May 1906. Salvage was attempted but it was only possible to remove her guns; the salvage was then abandoned.

Albemarle Mediterranean Fleet 1903-05. Thereafter Channel, Home and Atlantic Fleets until 1913. Joined 6 BS in the Grand Fleet 1914 and then 3 BS Channel Fleet. In November 1915 she had sailed, en route to the Mediterranean, when she suffered considerable weather damage in the Pentland Firth and had to return for repairs. In 1916 she went to Archangel as an icebreaker and guardship. In 1917 she returned home to go into Reserve. Sold 1920.

Cornwallis Mediterranean Fleet 1904-05, then Channel and Atlantic Fleets until 1910 when she returned to the Mediterranean. Home Fleet 1912. Joined 6 BS 1914 but was detached early in 1915 to join the Dardanelles force, where she played a major role in all the Dardanelles operations. Sunk by a U-boat off Malta 9 January 1917.

Below:
Duncan. *MPL*

Swiftsure and *Triumph*

Displacement: 11,800 (11,990)
Length: 480 (147)
Beam: 71 (21.6)
Draught: 25 (7.6)
Armour: 7in (17.8cm) belt, 10in (25.4cm) turrets
Armament: 4×10in; 14×7.5in; 14×14pdr; 2×12pdr; 2×18in torpedo tubes
Machinery: Twin screw, steam triple expansion
Fuel: Coal
Radius: 6,250nm
Speed: 19kt
Complement: 802

Unit	Laid Down	Completed	Builder
Swiftsure	Feb 1902	Jun 1904	Elswick
Triumph	Feb 1902	Jun 1904	Barrow

Design Notes
Both these ships were originally ordered by the Chilean government but were then bought by the British Admiralty. Although similar in many respects to contemporary British designs, they were somewhat longer and narrower than other battleships of the time. This gave them an advantage in speed, both ships exceeding 20kt on trials, the first battleships to do so, although their normal maximum was 19kt. Their armament was very different from other ships, having the 10in gun as the main armament rather than 12in and with the unusual size of 7.5in for the secondary battery. Their armoured protection was somewhat lighter than other battleships of the time and overall they were considered as 'lightweight' battleships. The goose-neck cranes either side of the funnel were easily-recognised identification features.

Historic Notes
Swiftsure Served in home waters until sent to the Mediterranean 1909-12. After a refit at home, she went out to the East Indies as Flagship in 1913. In 1914 she was used on operations in the Red Sea/Suez Canal area before joining the Dardanelles force in March 1915. At the end of that campaign in early 1916 she returned home to go into Reserve. Used as a gunnery target 1919. Sold 1920.

Triumph Served in Channel and Home Fleet 1904-09. Mediterranean Fleet 1909-12, then after a brief period at home went out to the China station in 1913. In 1914 took part in operations with Japanese ships before deploying to the Mediterranean to join the Dardanelles force in early 1915. Sunk by torpedo attack from a submarine whilst engaged in a bombardment off Gaba Tepe in the Dardanelles 25 May 1915.

Below:
***Swiftsure*, probably on sea trials 1904.** *Vickers*

Right:
***Swiftsure* in the Suez Canal 1915. The goose-neck cranes were unique to *Swiftsure* and *Triumph*.** *MPL*

Above:
***Dominion*. One of the single 9.2in mountings can be seen.**
Vickers Shipbuilding & Engineering Ltd

'King Edward VII' Class

Displacement: 16,350 (16,610)
Length: 453 (139)
Beam: 78 (23.8)
Draught: 27 (8.3)
Armour: 9in (22.8cm) belt, 12in (30.5cm) turrets
Armament: 4×12in; 4×9.2in; 10×6in; 14×12pdr; 4×18 torpedo tubes
Machinery: Twin screw, four-cylinder triple expansion steam
Fuel: Coal and oil
Radius: 7,000nm
Speed: 19kt
Complement: 777

Unit	Laid Down	Completed	Builder
King Edward VII	Mar 1902	Feb 1905	Devonport
Commonwealth	Jun 1902	Mar 1905	Fairfield
New Zealand	Feb 1903	Jun 1905	Portsmouth
Dominion	May 1902	Jul 1905	Vickers
Hindustan	Oct 1902	Jul 1905	Clydebank
Africa	Jan 1904	Nov 1906	Chatham
Britannia	Feb 1904	Sep 1906	Portsmouth
Hibernia	Jan 1904	Jan 1907	Devonport

Design Notes
Although part of a steady progression in battleship design, this class represented another step forward in armament and armoured protection: the 9.2in guns were a significant improvement in secondary armament. Although the ships' armour was generally thicker than in the 'Londons' and 'Duncans', improvements in armour-piercing (AP) shells were already making such armour less effective.

Historic Notes
At the beginning of World War 1, these ships formed the Third Battle Squadron (3 BS) with the Grand Fleet. However by early 1916 3 BS had been moved to the Thames Estuary to assist in the protection of the east coast of England.

King Edward VII Served in home waters 1905-12, being variously a Flagship in the Atlantic, Channel and Home Fleets. In 1912 deployed briefly to the Mediterranean; 3 BS 1914. On 6 January 1916 struck a mine off Cape Wrath and sank.

Commonwealth 1905-09 Atlantic Fleet, thence Home Fleet until 1914. 3 BS 1914-17; refit 1918. Gunnery ship at Invergordon 1919-21 after which she was scrapped

New Zealand (renamed *Zealandia* December 1911) 1905-13 served with Atlantic, Channel and Home Fleets. 3 BS 1914. Joined the Dardanelles force late 1915 and remained in the Mediterranean at the end of that campaign. Refit 1918, thence used as a depot ship before being sold in 1921.

Right:
King Edward VII. *Real Photos*

Dominion Atlantic Fleet 1905-09, thence to Home Fleet. 1914 3 BS. 1917-18 employed as a depot ship. Sold 1921.

Hindustan 1905-13 Atlantic, Channel and Home Fleets. 1914 3 BS, with which she remained throughout the war. Sold 1921.

Africa 1906-13 Atlantic, Channel and Home Fleets. In 1912 was fitted with a temporary flight deck over the forward turret for seaplane trials. 1914 3 BS. Joined 9th Cruiser Squadron in April 1917 until the end of the war. Then used as an accommodation ship until being sold 1920.

Britannia 1906-13 Atlantic, Home and Channel Fleets. 3 BS 1914. Was torpedoed off Cape Trafalgar two days before the end of the war on 9 November 1918.

Hibernia 1907-13 Atlantic, Channel and Home Fleets. In early 1912 a short flight deck was built over her forward turret and on 4 May 1912 Commander Samson made the first seaplane flight from a British warship under way. 3 BS 1914. Joined Dardanelles force November 1915 and remained in Mediterranean thereafter. Became accommodation ship 1919 and was sold in 1921.

'Lord Nelson' Class

Displacement: 16,500 (16,765)
Length: 443 (136)
Beam: 79 (24.0)
Draught: 27 (8.3)
Armour: 12in (30.5cm) belt, 12in (30.5cm) turrets
Armament: 4×12in; 10×9.2in; 24×12pdr; 5×18in torpedo tubes
Machinery: Twin screw, steam triple expansion
Fuel: Coal and oil
Radius: 9,000nm
Speed: 18kt
Complement: 800

Unit	Laid Down	Completed	Builder
Agamemnon	May 1905	Jun 1908	Beardmores
Lord Nelson	May 1905	Oct 1908	Palmers

Left:
***Lord Nelson* after 1917 when her funnels had been raised. The unusual arrangement of having five 9.2in guns in two twin and one single turret along the ship's side was unique to this class.** *Real Photos*

Design Notes
Although they were actually completed after *Dreadnought,* these two ships represent the last of the pre-dreadnoughts. However, they differed in several important respects from the earlier pre-dreadnoughts. They no longer mounted the 6in gun but had, in effect, a mixed main armament of 12in and 9.2in guns: at the same time, they were the last battleships to have a mixed main armament. They had an unusually built-up superstructure with the 12pdrs mounted high in it, firing over the 9.2s. They were also the last battleships to have reciprocating engines.

Historic Notes
Agamemnon 1908-13 Home Fleet. 1914 joined 5 BS, Channel Fleet. Joined the Dardanelles force in early 1915 and remained throughout that campaign. Thereafter she continued in the Mediterranean until the end of the war. She remained in Chatham 1919-23 when she was commissioned as a radio-controlled target ship for gunnery and aerial bombing. In one series of bombing trials, she remained unscathed after 114 bombs had been dropped!

Lord Nelson 1908-13 Home Fleet. 1914 Flagship of the Channel Fleet. Joined the Dardanelles force early in 1915 and remained in the Mediterranean for the rest of the war. Scrapped 1920.

IV
Dreadnought 1906

The advent of *Dreadnought* marked a significant step forward in battleship development. Such was the impact of her appearance that battleships built in the decade before became known as pre-Dreadnoughts and the classes which followed *Dreadnought* herself were known under the general name of 'Dreadnoughts'. She was very much the result of ideas put forward by Admiral 'Jackie' Fisher, although credit must go to Italian and American designers who were working along similar lines.

Dreadnought was primarily a result of the improvements in gunnery. In the early 1900s exercises and trials began to show that fighting ranges were likely to be of the order of 8,000yd as opposed to 2,000-3,000yd which was the current doctrine. The torpedo threat was one factor which led to the preference for longer ranges: if enemy battleships fired torpedoes, then one was safely out of range at 8,000yd. Equally, if the threat came from destroyers, then those destroyers would have to advance a long way outside the immediate protection of their own fleet in order to attack. At ranges of the order of 8,000yd, guns of the size 6in-9in were of limited effectiveness in inflicting damage upon a well-armoured battleship.

The requirement therefore was to obtain damaging hits with heavy guns at ranges of the order of 8,000yd. A reasonable chance of success could only be achieved by the use of salvo fire, ie a number of guns firing together. Salvo fire had two advantages. Firstly, the more shells were fired at the enemy ship, the greater were the chances of obtaining a single hit. Secondly, salvo fire allowed the fall of shot to be observed and corrections made for any error in range. Therefore, if one salvo fell short, the next could be fired with a slightly greater range set and adjustment continued until straddles were achieved, ie the shells falling in front of and beyond the target, with a high probability that one or more would hit. Effective salvo fire required as many guns as

possible to be fired together and for all the guns to be of the same size. Thus it was recognised that the future battleship should carry a main armament of a single size only and that as many guns as possible of that size should be mounted.

Having decided on the requirements for the main armament, the secondary armament was decided purely on the basis of defence against destroyers. Thus 12pdrs were considered adequate.

A number of other features were considered important in this next stage of battleship development. Speed was recognised as being a vital attribute in any engagement. Superior speed meant that an enemy could be chased and brought to battle, and perhaps more importantly also gave one the ability to control the range at which the action was fought and so maintain the tactical advantage. Therefore the bold decision was taken to adopt steam turbine propulsion. It had never been tried in a ship of the size of a battleship but turbines offered the advantages of increased power — and thus speed — improved reliability, the ability to maintain high speed for long periods, and a reduction in the overall weight of machinery.

When it came to deciding the armoured protection required, a degree of compromise was adopted. Comprehensive armoured protection would have meant an unacceptable increase in displacement and a reduction in speed. The turrets, barbettes and conning tower were of course well protected: the main sacrifice was made in the armoured belt along the ship's side. The height of this belt was reduced, the argument being that at the ranges contemplated for gunnery action, even out to 8,000yd, the trajectory of shells would still be almost flat. Thus the requirement to protect against plunging fire was not of paramount importance at the time that the *Dreadnought* plans were drawn up. (This was certainly not the case during *Dreadnought's* lifetime — plunging fire was to become a major concern.)

As part of the progress of providing adequate protection, important advances were made in the field of minimising the effects of damage should hits penetrate the armour. Very comprehensive water-tight sub-division was adopted with solid bulkheads to a point well above the waterline. Water-tight doors, seen as an area of weakness, were not fitted, and so movement from one water-tight compartment to another necessitated climbing up, along and down again. One major advance that made such a design possible was the use of oil fuel during action. The requirement to carry coal still existed mainly because coal supplies were easier to safeguard, Britain being a major coal-mining country, whereas the oil industry was very much in its infancy and the oil fields then known lay some considerable distance from the British Isles. However, the disadvantage of using coal during action was that it meant passages had to be kept open between the coal bunkers and the boiler rooms, thus reducing the effectiveness of the water-tight sub-division.

Various designs for battleships meeting these requirements were assessed. Superimposed turrets were considered, ie one turret mounted behind and higher than another so that one can fire over the top of the other. The Americans were already working on such a design (USS *Michigan*) but in British circles there were doubts over the vulnerability of the lower turret to blast damage from the higher. Finally a design which had all the main armament at the same level was adopted.

Some other finer points such as mounting the control position high on a mast which itself could withstand damage without collapsing were also incorporated. Hence a tripod mast which would stand if one leg was shot away was adopted.

From the combination of all these design developments, some of which were natural progressions from what had gone before and some of which were significant and bold changes — eg armament and main engines — emerged a design which was to have a great impact on the navies of the world.

Left:
***Dreadnought* in 1907 when Flagship of the Home Fleet.**
IWM

Dreadnought

Displacement: 17,900 (18,185)
Length: 527 (161)
Beam: 82 (25.0)
Draught: 26 (8.0)
Armour: 11in (28.0cm) belt, 11in (28.0cm) turrets
Armament: 10×12in; 27×12pdr; 5×18in torpedo tubes
Machinery: 4 screw, steam turbine
Fuel: Coal and oil
Radius: 6,600nm
Speed: 21kt
Complement: 730

Unit	Laid Down	Completed	Builder
Dreadnought	Oct 1905	Oct 1906	Portsmouth

Design Notes

The design incorporated all the major improvements discussed in this section. In addition her bows and forecastle were made higher than in earlier designs to improve sea-keeping. Her guns were originally destined for *Lord Nelson* and *Agamemnon* but were taken over in the interests of getting *Dreadnought* built in the extraordinary time of one year and a day. Her turbine main machinery was a great success. She was the first battleship to have four screws and twin rudders, the rudders being in line with the inboard screws. A final novelty in her design was that the positioning of the accommodation was reversed from the previous practice, *Dreadnought* having the officers' accommodation forward and that for the ratings aft. Although not perfect, she came up to all expectations and was a model for future designs for some years to come.

Historic Notes

After the remarkably short time in which she was built, *Dreadnought* carried out gunnery trials in late 1906. She then joined the Home Fleet as Flagship 1907-12. She was relieved as Flagship in May 1912, but stayed in the Home Fleet, becoming Flagship of 4 BS in late 1912, which she remained until 1916. She rammed and sank *U-29* in the North Sea on 18 March 1915. Was in refit at the time of Jutland. After the bombardment of Lowestoft by German battlecruisers in April 1916, *Dreadnought*, by then considered too slow for the Grand Fleet, joined 3 BS based in the Thames Estuary. She went into Reserve in 1919 and was scrapped in 1922.

Below:
***Dreadnought* with torpedo net booms fitted.** *Real Photos*

V
The First Battlecruisers 1908

The battlecruisers were very much another product of Admiral Fisher's ideas. He wanted ships which could take on the fast armoured cruisers which were entering service in the navies of the world. For this he required a class of exceptional ships which could both out-gun armoured cruisers and which were also faster than the fastest cruiser: their main armament would be almost the same as *Dreadnoughts*, 8×12in guns as opposed to the latter's 10×12in. In order to achieve the necessary speed from a hull large enough to mount the guns, very powerful machinery would be required, the horsepower being almost double that of *Dreadnought*. To meet these requirements, some sacrifice had to made and that sacrifice was in protection.

And so there emerged a new type of ship. The battlecruiser was to be armed to battleship standards but with the speed of a cruiser, and would thus have very powerful machinery, with, of necessity, large engine and boiler rooms. The armoured protection would be considerably lighter than that of a battleship. A number of classes of such ships were to be built over the years. They were ideal for situations such as the Battle of the Falkland Islands where only their superior speed allowed them to bring the German ships to battle. However, as part of the battle fleet at Jutland, they were very vulnerable: speed rather than armour was their protection and when employed with battleships they could not use their speed.

In the short space of two years, two revolutionary types of ship emerged on the world scene. The first had been *Dreadnought*, and then in 1908 the 'Invincible' class was revealed to the public.

Below:
Inflexible. *Real Photos*

'Invincible' Class

Displacement: 17,250 (17,525)
Length: 567 (173)
Beam: 78 (23.8)
Draught: 27 (7.3)
Armour: 6in (15.2cm) belt, 7in (17.8cm) turrets
Armament: 8×12in; 16×4in; 5×18in torpedo tubes
Machinery: 4 screw, steam turbines
Fuel: Coal and oil
Speed: 28kt
Complement: 784

Unit	Laid Down	Completed	Builder
Invincible	Apr 1906	Mar 1908	Elswick
Indomitable	Mar 1906	Jun 1908	Fairfield
Inflexible	Feb 1906	Oct 1908	Clydebank

Design Notes
These ships met Fisher's requirements for guns and speed, although the diagonal arrangement of the two midships turrets meant that these turrets could not fire across the ship, thus limiting the broadside to six guns. All the ships reached 28kt in service and *Indomitable* crossed the Atlantic at an average speed of 25kt. It was in armoured protection that these ships were deficient: in particular the turrets and magazines were vulnerable to plunging fire as a descending shell could explode within the barbettes, with little to prevent the flash going down to the magazine to cause a massive internal explosion.

Historic Notes
Indomitable After commissioning in 1908, the ship took HRH Prince George to the Canadian Tercentenary celebrations. Thence served in the Home Fleet, mainly as Rear Admiral Flagship 1st Cruiser Squadron until 1912. In early 1913 joined the 1st Battle Cruiser Squadron (1 BCS) but in late 1913 went out to the Mediterranean to join 2 BCS. At the beginning of World War 1 took part in the hunt for the *Goeben* and *Breslau*, and then in the Dardanelles blockade to the end of 1914. In early 1915 she returned home for a short refit and then joined the Battle Cruiser force in the Grand Fleet. Took part in the Dogger Bank action when she towed *Lion* home. She was part of 3 BCS for Jutland which she survived undamaged, and then joined 2 BCS for the remainder of the war. Went into Reserve in 1919 and was sold 1922.

Invincible Served with the Home Fleet in 1 CS 1909-13, thence joined 1 BCS until August 1913 when she went out to the Mediterranean until the end of the year. At the beginning of the war she served on the East Coast and took part in the Heligoland Bight action of 28 August 1914. In November 1914 she was detached to the South Atlantic and took part in the Battle of the Falkland Islands. She returned to the Grand Fleet in May 1915. At the Battle of Jutland she was hit in Q turret which caused a magazine explosion, broke in half and sank immediately, only five men being saved.

Inflexible Served with the Home Fleet 1909-12 and then went out to the Mediterranean as Flagship. Early in World War 1 she took part in the hunt for the *Goeben* and *Breslau* in the Mediterranean and then returned for duties on the East Coast briefly. In November 1914 she was detached to the South Atlantic and took part in the Battle of the Falkland Islands. Returned to the Mediterranean after this action and took part in the Dardanelles campaign. In March 1915 she was heavily damaged by a mine and had to retire to Gibraltar for repairs. Joined the Grand Fleet mid-1915 where she remained until the end of the war, taking part in Jutland without suffering any damage. After the war she went into Reserve and was sold in 1922.

Below:
***Invincible*. The long, low lines were distinctive features of the early battlecruisers.** *MPL*

VI
Dreadnought to *Queen Elizabeth* 1909-14

This short period saw the introduction into service of a total of 30 battleships and battlecruisers. The growing threat from Germany had been recognised and so ambitious building programmes were undertaken, and the emphasis of the Fleet moved to the North Sea. At this time the nature of the Fleet was very much under the influence of one man, Admiral 'Jackie' Fisher, who had pressed through the original *Dreadnought* and battlecruiser concepts.

This period saw steady improvements in the quality of the ships. Armament increased from the 12in to the 13.5in gun as the main armament and, somewhat later, the 6in replaced the 4in as the secondary armament. Gunnery ranges increased although ships were still being armoured with a fighting range of around 9,000yd in mind whereas many officers believed that 14,000-15,000yd was realistic — the important difference was that the shorter ranges did not demand armoured protection from plunging fire. Some ships were to be found sadly lacking in this protection during the war. Gunnery also improved with the introduction of director control, with the guns under the centralised control of a director mounted high in the ship. The high viewpoint improved spotting and range finding and made effective salvo fire at longer ranges a realistic proposition. Fighting ranges did, in fact, almost double in this period.

Propulsion improved steadily with the battlecruisers being fitted with the most powerful engines supported by a large number of boilers (42 in the 'Lion' class). In fact, because of the increased hull size required to carry the engines and boilers, the battlecruisers became the largest and heaviest ships of their time. They were some 6-8kt faster than the battleships with a similar gun armament, but much lighter armour.

The ships built in this period formed the Grand Fleet in 1914. For most of them their only moment of glory was the Battle of Jutland and many of them were rapidly discarded after the war under the terms of the Washington Treaty. Only the 'King George V' and subsequent classes were to see active service after the war.

Below:
***Bellerophon*. Two 4in guns can be seen mounted on top of the forward turret. The torpedo net is rolled up along the ship's side.** *Real Photos*

'Bellerophon' Class

Above right:
***Superb*, probably about 1910.** *MPL*

Displacement: 18,600 (18,895)
Length: 526 (161)
Beam: 82 (24.6)
Draught: 27 (8.3)
Armour: 10in (25.4cm) belt, 11in (28.0cm) turrets
Armament: 10×12in; 16×4in; 3×18in torpedo tubes
Machinery: 4 screw, steam turbine
Fuel: Coal and oil
Radius: 5,720
Speed: 21kt
Complement: 733

Unit	Laid Down	Completed	Builder
Bellerophon	Dec 1906	Feb 1909	Portsmouth
Superb	Feb 1907	May 1909	Elswick
Temeraire	Jan 1907	May 1909	Devonport

Design Notes
These ships were effectively repeat *Dreadnought*s with some improvements. Their secondary armament was increased from 12pdr to 4in, and some changes were made to the layout of the armoured protection, in particular with respect to defence against torpedoes. Also the layout of the masts was changed, the positioning of a mast with a control top immediately aft of the forward funnel having been found most unsatisfactory in *Dreadnought*.

Historic Notes
Bellerophon 1909-14 served entirely with the Home Fleet. Throughout the war was part of 4 BS, Grand Fleet; and survived Jutland undamaged. Became a gunnery drill ship after the war. Sold 1921.

Superb Home Fleet 1909-14. 1 BS then 4 BS Grand Fleet during the war. No damage at Jutland. In 1918 went out to Mediterranean and led the fleet through the Dardanelles after the surrender of Turkey. In 1919 became a gunnery drill ship. Sold 1921.

Temeraire 1909-14 Home Fleet. 4 BS Grand Fleet 1914-18. Undamaged at Jutland. Went to Mediterranean in 1918 and became a training ship in 1919. Sold 1922.

'St Vincent' Class

Displacement: 19,250 (19,545)
Length: 536 (164)
Beam: 84 (25.6)
Draught: 28 (8.6)
Armour: 10in (25.4cm) belt, 11in (28.0cm) turrets
Armament: 10×12in; 24×4in; 3×18in torpedo tubes
Machinery: 4 screw, steam turbine
Fuel: Coal and oil
Radius: 6,900nm
Speed: 21kt
Complement: 758

Unit	Laid Down	Completed	Builder
Vanguard	Apr 1908	Feb 1910	Vickers
Collingwood	Feb 1908	Apr 1910	Devonport
St Vincent	Dec 1907	Apr 1910	Portsmouth

Design Notes
These ships were almost identical in outward appearance to the 'Bellerophons' but, in fact, they were slightly larger. They also mounted the 50 calibre 12in gun as opposed to the 45 calibre which had been fitted in earlier classes. This gave a main armament with a higher muzzle velocity and thus greater penetrating power. The layout of the armour was also modified — less protection was provided for the bow and stern sections.

Historic Notes
Vanguard 1910-14 Home Fleet. 1914 onwards served with the Grand Fleet. Suffered no damage at Jutland, but was destroyed by an internal explosion at Scapa Flow 9 July 1917 with a large loss of life.

Collingwood 1910-14 Home Fleet. Grand Fleet throughout the war — the future King George VI served onboard as a lieutenant. No damage at Jutland. In 1919 went into Reserve as a gunnery training ship. Sold 1922.

St Vincent 1910-14 Home Fleet; Grand Fleet 1914-18 including Jutland (no damage). Into Reserve 1919 until sold 1922.

Centre right:
***Vanguard*. Despite the large number of 4in guns fitted, where they are is not obvious. Two are mounted on the after turret, two which would normally be on the forward turret are not fitted here, there is one beneath the forward funnel, two on top of the turret beneath the mainmast and three mounted in a battery beneath the after funnel.**
Vickers Shipbuilding & Engineering Ltd

Right:
***St Vincent*. The three 4in guns in the superstructure beneath the after funnel can just be seen.** *MPL*

Neptune

Displacement: 19,900 (20,220)
Length: 546 (167)
Beam: 85 (26.0)
Draught: 28 (8.6)
Armour: 10in (25.4cm) belt, 11in (28.0cm) turrets
Armament: 10×12in; 16×4in; 3×18in torpedo tubes
Machinery: 4 screw, steam turbine
Fuel: Coal and oil
Speed: 21kt
Complement: 759

Unit	Laid Down	Completed	Builder
Neptune	Jan 1909	Jan 1911	Portsmouth

Design Notes
Neptune was the result of a requirement to increase the number of guns which could be fired on the broadside: in the *Dreadnought* layout, the two midships turrets could only be fired on their respective sides. In *Neptune* the midships turrets were staggered and breaks were incorporated in the superstructure so that these turrets could fire across the ship. In order to accommodate this arrangement without too large an increase in overall length, the after turrets were superimposed, the first time this arrangement was used in a British battleship.

She was also the first battleship to have cruising turbines which reduced fuel consumption when steaming at moderate speed, and the first to be fitted with director control of the main armament when this system was introduced in 1912.

Historic Notes
Home Fleet 1911-14, including a period as Flagship. Was with the Grand Fleet throughout the war, and in action at Jutland — no damage. Went into Reserve 1919 and was sold in 1922.

Below:
***Neptune*. Her superimposed after turrets were the first in a British battleship; the small gunnery director can be seen on the foremast beneath the control top.** *IWM*

'Indefatigable' Class

Displacement: 18,800 (19,100)
Length: 590 (180)
Beam: 80 (24.4)
Draught: 27 (8.3)
Armour: 6in (15.2cm) belt, 7in (17.8cm) turrets
Armament: 8×12in; 16×4in; 2×18in torpedo tubes
Machinery: 4 screw, steam turbine
Fuel: Coal and oil
Speed: 25kt
Complement: 800

Unit	Laid Down	Completed	Builder
Indefatigable	Feb 1909	Feb 1911	Devonport
New Zealand	Jun 1910	Nov 1912	Fairfield
Australia	Jun 1910	Jun 1913	John Brown

Design Notes

This was the second class of battlecruisers, and they had many similarities to the 'Invincibles'. However, they mounted the 50 calibre 12in gun as opposed to the 45 calibre gun and the midships turrets were staggered to a greater extent than in the 'Invincibles' to improve the arcs of fire. The armoured protection was again weak, a factor which was to lead to the loss of *Indefatigable* at Jutland.

Historic Notes

Indefatigable Home Fleet 1911-13. In late 1913 joined 2 BCS in the Mediterranean. Took part in the search for *Goeben* and *Breslau* early in the war and thence was involved in the initial stages of the Dardanelles blockade. In early 1915 had a short refit in Malta then joined 2 BCS in the Grand Fleet. In action at *Jutland* during which she engaged the German battlecruiser *Von der Tann*. *Indefatigable* was hit several times and it is believed one or more shots penetrated to her magazines whose explosion sank her.

Australia After completion in 1913 served with the Australian Navy. At the beginning of World War 1 was involved initially with operations in the South Atlantic and Pacific, sinking the German merchant ship *Eleonore Woermann* near the Falkland Islands on 6 January 1915. Thereafter she joined the Grand Fleet, but missed Jutland while under repair after a collision with *New Zealand*. Thence was Flagship of 2 BCS until the end of the war. She then returned to the Australian Navy as Flagship but was scuttled off Sydney as part of the force reductions resulting from the Washington Treaty in 1924.

New Zealand Was built at the expense of the New Zealand government but was a Royal Navy ship. In early 1913 she commenced a world tour which took her to many parts of the then British Empire. At the end of 1913 she visited the Russian Baltic ports. In 1914 she joined the Grand Fleet and was in action against the *Blucher* at Dogger Bank. At Jutland she received one hit on one of her turrets. She completed the war with the Grand Fleet. After the war she took Admiral Sir John Jellicoe on a world tour for his study into future defence requirements. Sold 1922.

Above:
***Indefatigable* as built.** *MPL*

Below:
***New Zealand*. The bridge structure and arrangement of directors and control tops on the foremast has been considerably modified compared with *Indefatigable*.** *MPL*

'Colossus' Class

Displacement: 20,000 (20,320)
Length: 546 (167)
Beam: 85 (26.0)
Draught: 29 (8.9)
Armour: 11in (28.0cm) belt, 11in (28.0cm) turrets
Armament: 10×12in; 16×4in; 3×21in torpedo tube
Machinery: 4 screw, steam turbine
Fuel: Coal and oil
Radius: 6,700nm
Speed: 21kt
Complement: 755

Unit	Laid Down	Completed	Builder
Colossus	Jul 1909	Jul 1911	Scotts
Hercules	Jul 1909	Aug 1911	Palmers

Design Notes
These two ships were almost repeat *Neptunes*. The main differences were a modified arrangement of the secondary armament, some changes in armoured protection, and the fitting of only one mast abaft the fore funnel. The reason for this arrangement is not clear, it having already been found to be unsatisfactory in *Dreadnought*. They were the first ships to be fitted with 21in torpedo tubes, which could carry a very powerful warhead. By this time the torpedo had developed considerably and could be used effectively out to 6,000-7,000yd, with speeds up to 45kt if used at shorter ranges.

Historic Notes
Colossus Home Fleet 1911-14. Served with the Grand Fleet throughout the war. Was the only battleship in the main Battle Fleet to be hit at Jutland, the main damage being from a shell which exploded just abaft the fore funnel on the port side. After the war she served briefly as a training ship before being sold in 1920.

Hercules Served with Home Fleet 1911-14 and thence with the Grand Fleet throughout the war. No damage at Jutland. After the war served as headquarters ship for the Armistice Commission at Kiel. Sold 1920.

'Orion' Class

Displacement: 22,500 (22,860)
Length: 581 (178)
Beam: 88 (26.9)
Draught: 29 (8.8)
Armour: 12in (30.0cm) belt, 11in (28.0cm) turrets
Armament: 10×13.5in; 16×4in; 3×21in torpedo tubes
Machinery: 4 screw, steam turbine
Fuel: Coal and oil
Speed: 21kt
Complement: 752

Unit	Laid Down	Completed	Builder
Orion	Nov 1909	Jan 1912	Portsmouth
Monarch	Apr 1910	Mar 1912	Armstrongs
Thunderer	Apr 1910	Jun 1912	Thames Iron Works
Conqueror	Apr 1910	Nov 1912	Beardmore

Design Notes
The major change with this class was a return to the 13.5in gun, the advantage being that the same or greater hitting power could be achieved for a lower muzzle velocity: the amount of wear in a gun barrel increases with greater velocity and thus the larger guns would have a longer barrel life. The 12in guns fired an 850lb (395kg) shell whilst the 13.5in a 1,250lb (566kg) one. These were the first ships since *Dreadnought* to mount their main armament all on the centreline and the first battleships to have superimposed turrets both forward and aft. Their armoured protection was improved over earlier classes by raising the height of the belt armour along the side, which offered greater protection against plunging fire. More powerful machinery was required to propel the heavier hull at 21kt although the horsepower (27,000) was still well below that achieved by the battlecruisers. Their superstructure was much lower than previous classes and their silhouette was more reminiscent of a light cruiser. These low lines belied a very powerful ship with a weight of broadside almost double that of *Dreadnought*.

Historic Notes
Orion Home Fleet 1912-14 and Grand Fleet throughout the war. In action at Jutland. Remained in active service after the war until being sold in 1922.

Monarch Home Fleet 1912-14 and Grand Fleet throughout the war. Present at Jutland. Taken out of service 1922 but was used as a target for aircraft bombing and cruiser and battleship gunnery 20 January 1925. Did not finally sink until engaged by *Revenge* at close range.

Thunderer Home Fleet 1912-14 and Grand Fleet thereafter for duration of war. Present at Jutland. After the war became a training ship until being sold in 1926.

Conqueror Home Fleet 1912-14 and Grand Fleet throughout the war. Present at Jutland. Sold 1922.

Above:
***Colossus* shortly after completion.** *Real Photos*

Below:
***Monarch* before the fitting of a gunnery director.** *Vickers*

HMS *Thunderer*

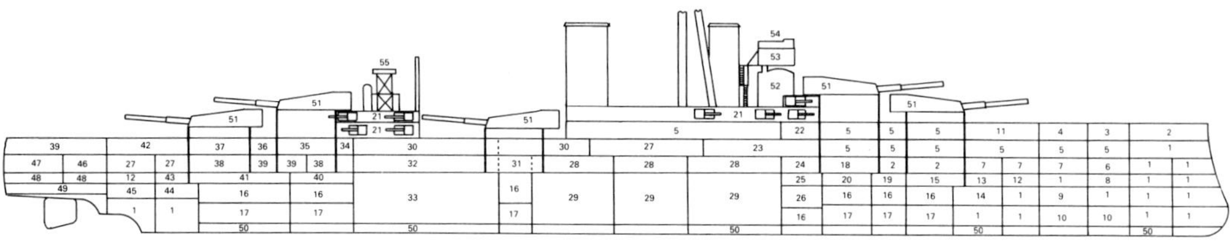

1 Watertight compartment
2 Officers' bathroom
3 Cabins
4 Wardroom pantry and galley
5 Cabins
6 Canvas and cordage room
7 Stores
8 Boatswain's store
9 Capstan engine room
10 Trimming tank
11 Wardroom
12 Freshwater tank
13 Pump room
14 Spirit room
15 Forward transmitting station
16 Magazine
17 12in shell room
18 Admiral's store
19 Switchboard room
20 Telephone exchange
21 4in battery
22 Admiral's cabin
23 Petty Officers' and Marines' messes
24 Electrical stores
25 Pump room
26 Torpedo room
27 Seamen's mess
28 Coal bunkers
29 Boiler room
30 Stokers' mess
31 Magazine cooling plant
32 Engineering workshops and stores
33 Engine room
34 Drying room
35 Stores
36 Canteen
37 Issue room
38 Medical distribution station
39 Ratings' bathroom
40 Hydraulic machinery
41 After transmitting station
42 Sickbay and Senior Ratings' mess
43 Flour store
44 Meat room
45 Hand steering compartment
46 Cells
47 Print room
48 Stores
49 After torpedo compartment
50 Double bottoms
51 Twin 12in turret
52 Conning tower
53 Bridge
54 Compass platform
55 Searchlight platform

Figure 3 — HMS *Thunderer* internal layout

Below:
***Orion*. The low lines of this class were more reminiscent of a light cruiser although with 10 of the new 13.5in guns they were very powerful ships.** *IWM*

'Lion' Class

Displacement: 26,350 (26,770)
Length: 700 (214)
Beam: 88 (26.9)
Draught: 29 (8.8)
Armour: 9in (22.9cm) belt, 9in (22.9cm) turrets
Armament: 8×13.5in; 16×4in; 2×21in torpedo tubes
Machinery: 4 screw, steam turbine
Fuel: Coal and oil
Radius: 5,610nm
Speed: 27kt
Complement: 997

Unit	Laid Down	Completed	Builder
Lion	Nov 1909	May 1912	Devonport
Princess Royal	May 1910	Nov 1912	Vickers
Queen Mary	Mar 1911	Sep 1913	Palmers

Design Notes
This class of battlecruiser represented enormous increases in length and displacement over even the 'Orions', the reason being the requirement for a ship which could carry an armament of 8×13.5in guns at 27kt. In some ways the design was unnecessarily complicated by retaining the midships turret rather than just having A, B, X and Y turrets; nevertheless the length was still required not only to give a hull capable of 27kt but also to carry the very powerful machinery required.

The shaft horse power was of the order of 70,000 compared with 27,000 in the 'Orions', requiring 42 boilers compared with 18 in the 'Orions'. Notwithstanding the fact that lighter armour was accepted for battlecruisers, the addition of the armoured protection then took the displacement to over 26,000 tons.

As built, the foremast was stepped immediately behind the fore funnel. When the ship was steaming hard, the heat from the funnel made the control top untenable, and so the mast was subsequently moved forward of the funnel. The ships enjoyed a fine reputation in the Service, often being reported as exceeding 30kt although 28½kt was the best ever achieved.

Historic Notes
Lion Flagship of 1 CS 1912 which became 1 BCS. During the war was Flagship of the Battle Cruiser Fleet throughout (Admirals Beatty and Pakenham). In action at Dogger Bank during which she received several hits and considerable damage to her engine rooms. She had to be towed home by *Indomitable* and then underwent four months of repairs. At Jutland a hit on Q turret ignited cordite charges, and if it had not been for the flooding of the magazine by the officer of the turret, the magazine would almost certainly have blown up with the loss of the ship. She remained Flagship of 1 BCS until discarded in 1922.

Princess Royal Served in 1 BCS throughout her career 1912-22. In action at Heligoland Bight and Dogger Bank without damage, but suffered several hits and considerable damage at Jutland. Discarded 1922.

Queen Mary 1 BCS throughout her short career 1913-16. In action at Heligoland Bight without damage, but at Jutland she was engaged by both the *Seydlitz* and *Derfflinger* at ranges of about 15,000yd. At this range protection was required against plunging fire, a feature in which this class was weak, and she was hit forward which caused a large explosion, followed by a second explosion amidships. She rolled over rapidly and sank with a large loss of life.

Below:
Princess Royal. *Real Photos*

Bottom:
***Lion*, probably about 1916.** *IWM*

'King George V' Class

Displacement: 23,000 (23,370)
Length: 597 (182)
Beam: 89 (27.2)
Draught: 29 (8.9)
Armour: 12in (30.5cm) belt, 11in (28.0cm) turrets
Armament: 10×13.5in; 16×4in; 3×21in torpedo tubes
Machinery: 4 screw, steam turbine
Fuel: Coal and oil
Speed: 21kt
Complement: 780

Unit	Laid Down	Completed	Builder
King George V	Jan 1911	Nov 1912	Portsmouth
Ajax	Feb 1911	Mar 1913	Scotts
Centurion	Jan 1911	May 1913	Devonport
Audacious	Feb 1911	Oct 1913	Lairds

Design Notes

These ships were effectively a repeat of the 'Orion' class with a slight increase in dimensions and some changes to the armoured protection. At the time they were considered under-gunned with respect to their secondary armament, the 4in gun being considered ineffective against the destroyers of the day which were now able to fire their torpedoes from longer range.

Historic Notes

King George V Home Fleet 1912-14 and thence Grand Fleet throughout the war. In action at Jutland, undamaged. Flagship Mediterranean Fleet 1919-23, thence into Reserve until being sold 1926.

Ajax Home Fleet 1913-14 thence with Grand Fleet throughout the war. In action at Jutland — no damage. 1919-24 Mediterranean Fleet, including operations in the Black Sea. In Reserve 1924-26 and then sold.

Centurion She had a long and mixed career. She sank an Italian merchant ship after a collision early in her trials. Home Fleet 1913-14 and thence Grand Fleet throughout the war including action at Jutland. She was in the Mediterranean 1919-24, including Black Sea operations, and then went into Reserve 1924-26. Was then converted to a radio controlled target ship and was used during gunnery practices. Remained thus until 1941 when she was disguised to look like the new battleship *Anson* then still building. She left UK in May 1941 and made her way to Bombay via the Cape. Remained there until May 1942 when she moved to Alexandria. Here Oerlikon guns were fitted and *Centurion* was then used as an anti-aircraft ship, mainly at the southern end of the Suez Canal. She returned to UK in May 1944 and ended her days being sunk as one of the blockships forming part of the Mulberry harbour for the Normandy landings.

Audacious Home Fleet 1913-14; Grand Fleet at outbreak of war. On 27 October 1914 hit a mine off Loch Swilly which caused flooding. Despite attempts at salvage, she sank after an internal explosion some hours later.

Below:
King George V **about 1919. Note the aircraft flying-off platform on B turret.** *MPL*

'Iron Duke' Class

Displacement: 25,000 (25,400)
Length: 623 (190)
Beam: 90 (27.5)
Draught: 29 (8.9)
Armour: 12in (30.5cm) belt, 11in (28.0cm) turrets
Armament: 10×13.5in; 12×6in; 2×3in AA; 4×21in torpedo tubes
Machinery: 4 screw, steam turbine
Fuel: Coal and oil
Speed: 21kt
Complement: 940

Unit	Laid Down	Completed	Builder
Iron Duke	Jan 1912	Mar 1914	Portsmouth
Marlborough	Jan 1912	Jun 1914	Devonport
Benbow	May 1912	Oct 1914	Beardmore
Emperor of India	May 1912	Nov 1914	Vickers

Design Notes
In general these ships were very similar to the 'King George Vs'. They were slightly longer and heavier but the main difference was that their secondary armament was increased to the 6in gun from the 4in fitted in earlier classes; they were also fitted with anti-aircraft guns. They were very good ships.

Historic Notes
Iron Duke She was the Flagship of the Home Fleet 1912-14 and then of the Grand Fleet (Admiral Sir John Jellicoe) until 1916. Flagship at Jutland — no damage sustained. She remained in the Grand Fleet throughout the war and then served in the Mediterranean 1919-26, including operations against the Bolsheviks in the Black Sea. Returned to the Atlantic Fleet 1926-29 and then became a training ship with reduced armament until 1939. At the beginning of World War 2 she moved to Scapa Flow and was used as a depot ship and also for defence, some of her guns still being operational. She was damaged by aircraft bombing and had to be bottomed, but continued in service until the end of the war. Sold 1946.

Marlborough Home Fleet 1912-14 and thence Grand Fleet throughout the war. At Jutland she was struck by a torpedo which caused flooding of the compartments immediately forward of 'A' boiler room. She continued in action with the Fleet and indeed scored hits on the *Konig* shortly afterwards. However, overnight 'A' boiler room bulkhead showed signs of giving way so *Marlborough* withdrew at reduced speed and eventually reached the Tyne for repairs. After the war she served in the Mediterranean and Black Sea 1919-26, and then with the Atlantic Fleet until 1929. Sold 1932.

Benbow Grand Fleet throughout the war. No damage at Jutland. Mediterranean 1919-26 and thence Atlantic Fleet until 1929. Sold 1931.

Emperor of India Grand Fleet throughout the war. Undamaged at Jutland. Mediterranean 1919-26 and Atlantic Fleet until 1929. Sold 1931.

Below:
***Emperor of India*. She has five 6in guns forward, and one aft beneath X turret.**
Vickers Shipbuilding & Engineering Ltd

Below:
Emperor of India. *Real Photos*

Bottom:
***Iron Duke*, Flagship of the Grand Fleet. She had all her 6in guns forward.** *Real Photos*

Erin

Displacement: 23,000 (23,370)
Length: 559 (171)
Beam: 91 (27.8)
Draught: 28 (8.6)
Armour: 12in (30.5cm) belt, 11in (28.0cm) turrets
Armament: 10×13.5in; 16×6in; 6×6pdr; 2×3in AA; 4×21in torpedo tubes
Machinery: 4 screw, steam turbine
Fuel: Coal and oil
Speed: 21kt
Complement: 1,070

Unit	Laid Down	Completed	Builder
Erin	Aug 1911	Aug 1914	Vickers

Design Notes
Erin was ordered as the *Rechad V* for the Turkish Navy, and was taken over by the RN for World War 1. She was very similar to the 'King George V' and 'Iron Duke' classes with some differences in dimensions. She was well up to British standards of design and was an important addition to the Fleet.

Above:
***Erin* as completed.**
Vickers Shipbuilding & Engineering Ltd

Historic Notes
Served with Grand Fleet throughout the war. No damage at Jutland. Went into Reserve 1919 and was scrapped 1921.

Agincourt

Displacement: 27,500 (27,940)
Length: 672 (206)
Beam: 89 (27.2)
Draught: 27 (8.3)
Armour: 9in (22.8cm) belt, 12in (30.5cm) turrets
Armament: 14×12in; 20×6in; 10×3in; 2×3in AA; 3×21in torpedo tubes
Machinery: 4 screw, steam turbine
Fuel: Coal and oil
Speed: 22kt
Complement: 1,115

Unit	Laid Down	Completed	Builder
Agincourt	Sep 1911	Aug 1914	Elswick

Design Notes
Ordered originally by Brazil, she was sold to Turkey during build to be completed as the *Sultan Osman I*. However, at the beginning of World War 1 she was taken over by Britain. A generally sound design, she was unique in having seven main turrets. In fact overall she was heavily armed with a large secondary battery as well.

Historic Notes
Grand Fleet throughout the war; no damage at Jutland. Used for experimental purposes after the war before being scrapped 1921.

Below:
***Agincourt* as completed. She was unique in having seven main armament turrets.** *Vickers*

Tiger

Displacement: 28,500 (28,955)
Length: 704 (215)
Beam: 90 (27.5)
Draught: 28 (8.6)
Armour: 9in (22.8cm) belt, 9in (22.8cm) turrets
Armament: 8×13.5in; 12×6in; 2×3in AA; 4×21in torpedo tubes
Machinery: 4 screw, steam turbine
Fuel: Coal and oil
Speed: 29kt
Complement: 1,120

Unit	Laid Down	Completed	Builder
Tiger	Jun 1912	Oct 1914	John Brown

Design Notes
A magnificent ship, she followed the 'Lion' class as the next battlecruiser. It may seem odd that the battlecruisers which were designed to be lightly armoured and fast were also the largest and heaviest ships of their day, but, as discussed with the 'Lion' class, this was mainly due to the space required for the boilers and machinery dictating a longer ship and thus a greater overall weight of armour. *Tiger* is often quoted as being able to exceed 30kt: this is not correct.

Above:
***Tiger* in July 1924.** *IWM*

Historic Notes
Served with the Grand Fleet throughout the war. In action at Dogger Bank where she was hit several times, including damage to B turret. She was hit 21 times at Jutland, including some on A, Q and X turrets, repairs after the battle taking two months. She joined the BCS, Atlantic Fleet 1919-22 and again 1929-31, serving as a gunnery training ship in the interim. Sold 1932.

Canada

Displacement: 28,000 (28,450)
Length: 661 (202)
Beam: 92 (28.1)
Draught: 29 (8.9)
Armour: 9in (22.9cm) belt, 10in (25.4cm) turrets
Armament: 10×14in; 16×6in; 2×3in AA; 4×21in torpedo tubes
Machinery: 4 screw, steam turbine
Fuel: Coal and oil
Speed: 22.75kt
Complement: 1,165

Unit	Laid Down	Completed	Builder
Canada	Dec 1911	Sep 1915	Elswick

Design Notes
Although completed after the 'Queen Elizabeths', *Canada* was laid down a year before the new generation of battleships was commenced and can therefore be said to be part of the pre-'Queen Elizabeth' era. She was ordered originally by Chile as the *Almirante Latorre* but was purchased by Great Britain. She was effectively a slightly enlarged 'Iron Duke' with 14in main armament rather than 13.5in guns. She was also a fast ship and therefore made a welcome addition to the Grand Fleet.

Historic Notes
Canada served as part of the Grand Fleet throughout the war, being undamaged at Jutland. After the war she was transferred to Chile where she served until 1960.

Below:
***Canada*, her superstructure dominated by her enormous after funnel.** *IWM*

VII
Battleships for Two World Wars 1915-17

During this short period were built 12 ships which were to see service in both world wars. All of them underwent modification and improvement between the wars and four of them were extensively rebuilt.

The 15in gun became the standard, originally mounted in all ships in the Mk 1 twin mounting. This gave a maximum elevation of 20deg allowing a 1,920lb (870kg) shell to be fired 28,700yd. The four reconstructed ships had their turrets modified to 30deg elevation which increased the range to 32,200yd. The secondary armament of all the ships on building was designed to counter the destroyer threat, and they were all re-armed with dual-purpose guns whose primary role was anti-aircraft but which could be used against surface targets as well. In addition, a considerable amount of smaller anti-aircraft guns were added, including 2pdr on multiple mounts — known universally as pom-poms — 0.5in machine guns and 20mm Oerlikons.

All the ships were lacking in armour when built, particularly deck armour, and eventually most of them had considerable amounts of deck armour added plus improvements in side armour and anti-torpedo bulging. After Jutland anti-flash protection was fitted to the magazines.

Machinery continued to improve and these 12 ships were faster for their type than their predecessors. All ships were fitted with flying-off platforms for aircraft either around the end of World War 1 or in the early 1920s, all of which were subsequently replaced by catapults. All ships also had their bridge areas improved and gunnery control arrangements modified.

Ships built in this period were to form the backbone of the battleship force in World War 2, in particular providing the bulk of the Mediterranean Fleet throughout the war in that area, the Eastern Fleet until 1944, and also providing convoy escorts, an unglamorous but essential task to counter the threat of surface raiders.

Below:
Ships of the 'Royal Sovereign' class, *Royal Sovereign* nearest. *Real Photos*

'Queen Elizabeth' Class

AS BUILT
Displacement: 27,500 (27,940)
Length: 645 (197)
Beam: 90 (27.5)
Draught: 31 (9.5)
Armour: 13in (33.0cm) belt, 13in (33.0cm) turrets, 1.75in (4.4cm) upper deck
Armament: 8×15in; 14×6in (16 in *Queen Elizabeth*); 2×3in AA; 4×21in torpedo tubes
Machinery: 4 screw, steam turbine
Fuel: Oil
Speed: 24kt
Complement: 950

Unit	Laid Down	Completed	Builder
Queen Elizabeth	Oct 1912	Jan 1915	Portsmouth
Warspite	Oct 1912	Mar 1915	Devonport
Barham	Feb 1913	Oct 1915	John Brown
Malaya	Oct 1913	Feb 1916	Armstrongs
Valiant	Jan 1913	Feb 1916	Fairfield

Design Notes
In many ways this class was the logical follow-on from *Iron Duke* and *Tiger* but some simple changes made the class a major step forward in battleship development. Firstly they mounted the new 15in gun. This 42 calibre gun was capable of firing a 1,920lb (870kg) shell 23,400yd, and thus 8×15in guns provided a very similar weight of broadside to 10×13.5in. The removal of Q turret from the design meant that space in the centre of the ship devoted, in previous classes, to magazines and shell rooms, could now be used for propulsion machinery. Effectively this made room for another boiler room and the 'Queen Elizabeths' had 24 boilers in four boiler rooms against 18 in three boiler rooms in the 'Iron Dukes', allowing an increase in speed to 24kt. They were also the first battleships to be all oil-fired.

The secondary armament was similar to the 'Iron Dukes', mounted in batteries stretching from B turret to the after funnel along the upper deck. This position was not very satisfactory in the 'Iron Dukes' as the batteries were liable to flooding in anything of a seaway, and so some improvements were made in the 'Queen Elizabeths'. In all but *Queen Elizabeth* herself the after-most gun was not mounted, hence the fact that most of the class only mounted 14×6in guns.

Armoured protection was also similar to the 'Iron Dukes', and all ships were fitted with main armament directors during build. They had particularly high metacentric heights and thus were very stable ships.

They were an extremely successful class. However, their ability to fight not only during World War 1 but also during World War 2 with considerable distinction was only possible through a comprehensive programme of modifications during their lives.

MODIFICATIONS
1916-20 Stiffening fitted in fore part of ship and improvements made to watertight integrity. Anti-flash arrangements fitted to magazines. Searchlight control positions added, and aircraft flying-off platforms were fitted on B and X turrets in 1918. Changes made to masts.

After these small changes, all the ships except *Barham* underwent two major modifications before World War 2. *Barham* was only modified once.

Below:
***Warspite* at Scapa Flow in 1916, showing the class as completed with two funnels and the 6in battery forward.**
IWM

1924-30 *Queen Elizabeth*, *Warspite*, *Malaya*, *Valiant*. Bulges were fitted along the waterline at the sides to improve the protection against torpedoes and to provide additional buoyancy in the event of damage. The 2×3in AA guns were replaced by 4in HA (high angle) mounts and a High Angle Control Position (HACP) was added. The bridge, control positions and tops were all modified, in particular to improve the arrangement of gunnery control positions. The twin funnels were trunked into a single funnel to solve problems of funnel gases around the bridge. There were other minor changes such as in wireless equipment, searchlights, etc. *Valiant* had an aircraft catapult fitted on the quarterdeck, pom-pom anti-aircraft guns mounted and her submerged torpedo tubes removed.

Barham. During this period only the change from 2×3in AA to 4×4in HA guns was made, along with the fitting of a temporary HACP.

1930-33 *Barham* underwent her only major modification. Bulges were fitted as in the other ships. An additional 4in (10.2cm) of deck armour was mounted over the magazines as further protection against plunging shells and bombs. Pom-pom mountings were fitted along with permanent HACPs for the 4in HA guns, and the after submerged torpedo tubes were removed. The aircraft launching platforms were removed and a single catapult fitted on X turret. The bridge and control positions were modified and the funnels were trunked.

1934-37 *Malaya* was modified for the second time along the same lines as *Barham*. Additional deck armour was fitted over both the magazines and the engine rooms, the four single 4in HA single mounts were replaced by twin mountings, and pom-pom AA guns were mounted abreast the funnel. The submerged torpedo tubes were removed. An aircraft catapult was fitted abaft the funnel, and the bridge area was remodelled and the conning tower reduced.

Warspite underwent major reconstruction. The 15in turrets were modified to give a maximum elevation of 30deg vice 20deg, and a new 15in director was fitted. The 6in battery was reduced to four guns each side and additional AA armament provided. Deck armour was placed over the magazines, engine and boiler rooms, and new boilers and engines were fitted, giving a much reduced fuel consumption and greater power. A midships aircraft catapult was mounted, while the bridge and superstructure were completely rebuilt.

1937-41 *Valiant* and *Queen Elizabeth* were reconstructed along similar lines to *Warspite*. The major additional change in these two ships was the removal of the 6in guns completely and the fitting instead of 20×4.5in in twin mountings.

World War 2 The significant changes made to all the class except *Barham* during the war were the fitting of radar and additional anti-aircraft guns and, in *Warspite*, the removal of the 6in guns in 1944.

Above:
***Valiant* after her first major modification. The twin funnels have been trunked into a single funnel, the bridge structure modified and, in this class in the case of *Valiant* only, an aircraft catapult fitted on the quarterdeck.** *Real Photos*

Below:
***Queen Elizabeth* in May 1937, after her first, but before the second, major modification.** *MPL*

THE CLASS IN WORLD WAR 2

The ships which fought in World War 2 were very different from those of World War 1. The main details of the class in 1943 compared with the 'as built' specification were:

Displacement: Increased by 3,000-4,000 tons
Length/beam: No change
Draught: Increased by about 6in (15cm)
Armour: Belt and turrets unchanged; deck armour increased to 5in (12.7cm)
Armament: *Queen Elizabeth/Valiant:* 8×15in (30deg elevation); Nil 6in; 20×4.5in; 24×2pdr (pom-pom); 16×0.5in
Warspite: 8×15in (30deg elevation); 8×6in; 8×4in; 32×2pdr; 31×20mm
Malaya: 8×15in (20deg elevation); 14×6in; 8×4in; 16×2pdr; 16×0.5in
Torpedo tubes removed in all ships
Machinery: *Queen Elizabeth, Valiant* and *Warspite* re-engined and re-boilered
Speed: *Queen Elizabeth* and *Valiant* 24kt; *Warspite* and *Malaya* 23.5kt

Note: Barham was sunk before 1943.

The other major changes had been the fitting of bulges, mounting of an aircraft catapult and, except *Malaya*, rebuilding of the bridge and superstructure.

Historic Notes

Queen Elizabeth On commissioning joined Mediterranean Fleet February 1915 and employed as Flagship in the Dardanelles, taking part in the major actions in March 1915. Joined Grand Fleet May 1915; absent from Jutland, but Flagship Grand Fleet 1916-18. German surrender signed on board November 1918. Thence Flagship Atlantic Fleet until 1924. Served in the Mediterranean almost without a break 1924-37. Refitted 1926-27; reconstruction 1937-January 1941. Joined Mediterranean Fleet May 1941; 19 December 1941 damaged by Italian 'human torpedo' attack Alexandria harbour, and after temporary repairs, proceeded to Norfolk, Virginia for permanent repair September 1942-June 1943. Returned to UK until December 1943 thence to Eastern Fleet, Trincomalee as Flagship January 1944. Operations along Burma and Malayan coasts. Returned to UK July 1945, into Reserve. Sold 1948.

Warspite Possibly the most famous British battleship. Joined Grand Fleet March 1915. Received 13 hits at Jutland, with considerable damage, including jammed steering gear (for a short period) and some flooding. After repair remained with Grand Fleet throughout the war. Atlantic Fleet 1919-24, thence refit until 1926. Mediterranean Fleet 1926-30, thence Atlantic/Home Fleet until 1934. Major reconstruction 1934-37. Mediterranean Fleet 1937-39. Rejoined Home Fleet early 1940. Troop convoy escort. Norwegian campaign April 1940; Second Battle of Narvik 13 April 1940. In May 1940 returned to Mediterranean as Flagship. Saw action off Calabria 9 July 1940 during which she obtained a hit on the Italian battleship *Giulio Cesare*. Shore bombardments including Bardia and Tripoli. 28 March 1941 Battle of Matapan. 22 May 1941 damaged by bomb during operations off Crete. After temporary repairs in Alexandria, proceeded via Singapore to Bremerton, Seattle, USA for repair. Joined Eastern Fleet January 1942 but returned to Mediterranean to take part in Salerno landings September 1943. During bombardment operations was severely damaged by two glider bombs which caused massive flooding of the boiler rooms. After temporary repairs in Malta and then Gibraltar, she arrived in Rosyth for further repair March 1944. Although not fully repaired, she took part in bombardment operations in support of the Normandy landings June 1944. Requiring a change of gun barrels after heavy wear, she sailed for Rosyth 11 June. On 13 June again severely damaged by a mine off Harwich; partial repair Rosyth. Rejoined bombardment force 25 August 1944; in action at Brest, Le Havre and Walcheren.

Below:
***Queen Elizabeth* after her second major modification, representing the final development of the class. The 6in battery has been removed and twin 4.5in mountings fitted on the upper deck.** *IWM*

Above:
***Malaya* with *Barham* in the background. These two ships were not modified as extensively as the other three of the class. The 6in battery has been retained and the twin 4in mountings can be seen on the upper deck amidships.** *IWM*

No further wartime operations. Whilst under tow to the breaker's yard 17 April 1947 she broke the tow and ran aground at Prussia Cove, Cornwall, where she was eventually broken up.

Barham Joined Grand Fleet October 1915; Flagship 5 BS at Jutland, hit six times. After repair rejoined Grand Fleet for remainder of war. Atlantic Fleet 1920-24, thence Mediterranean Fleet until 1927. Refitted 1927-28 and then returned to Mediterranean, until 1929. Atlantic Fleet 1929-30. Major refit 1930-33, thence service with Home Fleet until 1935. Returned to Mediterranean 1935-39. Returned to the Home Fleet at the beginning of the war. Damaged by U-boat torpedo attack 28 December 1939 off the Hebrides, and repaired in Liverpool. Operational again April 1940, supported the unsuccessful landings at Dakar 24 September 1940, and at Battle of Matapan 28 March 1941. Bombardment of Tripoli 21 April 1941. Operations in support of Crete evacuation May 1941: damaged by bombs. After temporary repair sent to Durban for permanent repair. Rejoined fleet July 1941, but torpedoed by *U-331* off Cyrenaica 25 November 1941: three torpedo hits port side between the funnel and Y turret, after which she quickly listed and started to turn over. As she turned on her beam ends, the after 15in magazine exploded, completely destroying the ship.

Malaya Her building was paid for by the Federated Malay States. Joined Grand Fleet February 1916. She sustained considerable damage from eight hits at Jutland, and rejoined fleet after repairs June 1916. With Grand Fleet for rest of the war. 1921 made a visit to Malaya, after which she joined the Home Fleet until 1924, although she was detached to the Mediterranean in 1922. Mediterranean Fleet 1924-27 and thence refit until 1929, followed by a return to the Mediterranean 1929-30. Atlantic/Home Fleet 1930-34. Refit 1934-36, thence back to the Mediterranean until the beginning of World War 2. For the first few months of the war she was engaged in escorting Atlantic convoys, and in March 1940 she rejoined the Mediterranean Fleet. Action off Calabria 9 July 1940. Supported convoy movements in the Mediterranean; in early 1941 returned to convoy escorting in the Atlantic, and on 20 March 1941 was torpedoed 250 miles west-northwest of the Cape Verde Islands, with some damage to her bulges and wing compartments. She went to the USA via Trinidad for repair. In July 1941 she returned to the Mediterranean and joined Force H covering Malta convoys. Continued mainly as a convoy escort until being placed in care and maintenance July 1943. She was recommissioned for bombardment in support of the Normandy landings but only carried out one operation, against the island of Cezembre. She went back in Reserve and was sold 1948.

Valiant Joined Grand Fleet February 1916, undamaged at Jutland. Atlantic Fleet 1919-24, then Mediterranean Fleet until 1929. Refit 1929-30 then Home/Atlantic Fleet 1930-35. Returned to Mediterranean 1935-37 and then underwent major reconstruction 1937-39. Was still in refit at outbreak of World War 2 but completed in December 1939. Went out to the West Indies to work up and then escorted a convoy from Halifax to UK to join the Home Fleet February 1940. Joined Force H for action against French at Mers El-Kebir 3 July 1940. Joined Mediterranean Fleet August 1940. Shore bombardments including Valona and Bardia, and at Battle of Matapan 28 March 1941. May 1941 took part in operations off Crete. 19 December 1941 was badly damaged by Italian 'human torpedo' attack in Alexandria harbour. Repaired in Durban April-July 1942 after which she joined the Eastern Fleet until February 1943 when she returned to UK for refit until May 1943. She then returned to the Mediterranean and covered the Sicily and Salerno landings. Rejoined Eastern Fleet January 1944, but was damaged in the floating dock at Trincomalee 8 August 1944 when the dock collapsed. This necessitated a return to Devonport for further repairs, still under way at the end of the war. The refit was completed in April 1946 when she became a training ship in Devonport. Sold 1948.

'Royal Sovereign' Class

AS BUILT
Displacement: 27,500 (27,940)
Length: 624 (191)
Beam: 102 (31.2)
Draught: 30 (9.2)
Armour: 13in (33.0cm) belt, 13in (33.0cm) turrets, 1½in (3.8cm) deck
Armament: 8×15in; 14×6in; 2×3in AA; 4×21in torpedo tubes
Machinery: 4 screw, steam turbine
Fuel: Oil
Speed: 21kt
Complement: 950

Unit	Laid Down	Completed	Builder
Revenge	Dec 1913	Mar 1916	Vickers
Royal Oak	Jan 1914	May 1916	Devonport
Royal Sovereign	Jan 1914	May 1916	Portsmouth
Resolution	Dec 1913	Dec 1916	Palmers
Ramillies	Nov 1913	Sep 1917	Beardmore

Design Notes
The particulars of this class are very similar to the 'Queen Elizabeths' except in one very important respect. They were designed for coal-fired boilers with lower-powered engines than the 'Queen Elizabeths', and although were changed to oil-fired boilers during build they were still limited in power. Therefore their speed of 21kt, adequate for World War 1, was insufficient to make them first-line ships in World War 2. Their armament was very similar to the 'Queen Elizabeths' except that the 6in battery was mounted farther back along the ship's side which made it less vulnerable to flooding in rough weather. They were fitted with main armament directors during build.

The armoured protection was similar to the 'Queen Elizabeths' although in some respects their deck armour was superior. *Ramillies* was fitted with torpedo bulges during build; the rest were fitted later. They were unusual for the period in only having one funnel which gave them a distinctive silhouette, and they were built with an unusual rudder arrangement — a single balanced centreline rudder with a smaller auxiliary rudder ahead of it. In practice the auxiliary rudder was unnecessary.

Like the 'Queen Elizabeths', the class went through a series of modifications up to World War 2 although the modifications carried out were not so extensive as in the 'Queen Elizabeths'.

MODIFICATIONS
1916-20 Watertight integrity and anti-flash arrangements for cordite hoists were improved. Stiffening fitted in forward part of ship, and 1in deck armour fitted over magazines. Secondary directors added and flying off platforms for aircraft mounted on B and X turrets. Bulges fitted to *Royal Sovereign*, *Revenge* and *Resolution*.

The modifications made between the wars were limited by a number of factors. The lack of speed appears to be one of them although new engines and boilers could have increased their speed, despite the lack of space for boiler rooms, only having three as opposed to the four in the 'Queen Elizabeths'. Also they did not lend themselves so readily to rebuilding of the superstrucure as the space available in the bridge area was reduced.

1921-28 Bulges were fitted to *Royal Oak*, the last of the class to be so fitted. The 3in AA guns were removed and replaced with 4in HA, the exact number varying from ship to ship, either two or four. In all but *Royal Oak* 2×6in guns were removed.

Below:
***Royal Oak* as completed in 1916. The 6in battery is farther aft than in the 'Queen Elizabeths'.** *IWM*

Above:
***Revenge* during World War 1.**
Vickers Shipbuilding & Engineering Ltd

1930-39 There were further changes to the 4in armament. Pom-pom and 0.5in close-range anti-aircraft guns were fitted. The flying off platforms on B and X turrets were removed and a single catapult fitted; deck armour was improved in *Royal Oak*. Submerged torpedo tubes were removed.
World War 2 The main changes were improvements to the AA armament, fitting of further deck armour and the fitting of radar.

THE CLASS IN WORLD WAR 2
The state of the class in 1942 (*Royal Oak* having been lost) compared with the 'as built' specification was:

Displacement: Increased by 2,000-3,000 tons
Length/beam: No change
Draught: Little change as the fitting of bulges appeared to balance the increase in displacement
Armour: Little change except for improvements in deck armour
Armament: 8×15in; 12×6in; 8×4in HA; 24×2pdr pom-poms; 10×20mm Oerlikons
Note: Some ships had additional 20mm added 1943-44.

Remainder of details unchanged.

Historic Notes
Revenge Joined the Grand Fleet March 1916. No damage at Jutland. Remained with the Grand Fleet until the end of the war, then served in the Mediterranean 1919-20. Thence service with the Atlantic and Mediterranean Fleets until 1931. Refit in 1925 and 1931, thence to Home Fleet outbreak of World War 2. Employed primarily on convoy escort duties 1939-41 although did take part in bombardment operations in the Channel in 1940. Joined Eastern Fleet early 1942 until September 1943, when she returned to UK. Reduced to a training role May 1944 and finally sold 1949.

Royal Oak Joined Grand Fleet May 1916 just in time for Jutland — no damage. Atlantic Fleet 1919-22 and 1924-26 with a refit in the interval. Mediterranean Fleet 1926-34, then refit. Home Fleet 1935. On the night of 13/14 October 1939 she was torpedoed in Scapa Flow by *U-47* (Lt Gunther Prien) and sank with a large loss of life.

Royal Sovereign Joined Grand Fleet May 1916 but missed Jutland. Remained with the Grand/Atlantic Fleets until 1926 then refit until 1928. Returned to Atlantic Fleet 1928-35. Reserve and training role 1936-37, thence to Home Fleet. Early in World War 2 was employed on escort duties in the Atlantic but then joined Mediterranean Fleet mid-1940. Action off Calabria 9 July 1940. In late 1941 she joined the Eastern Fleet. Refit in USA late 1943. In 1944 she was transferred to Soviet

Union and renamed *Archangelsk*. Returned to UK February 1949 and sold in the same year.

Resolution Joined Grand Fleet December 1916. Remained with Grand/Atlantic Fleet until refit 1924. Then served in Mediterranean 1925-35 apart from refit 1930-31; Home Fleet from 1936. Early in World War 2 was employed on convoy escort duties. In action during Norwegian campaign, and hit by bomb off Norway 16 May 1940 — slight damage. In action against French at Mers El-Kebir 3 July 1940, also supported operations off Dakar September 1940. Torpedoed by French submarine *Beveziers* 25 September 1940: considerable damage to bulge port side, and one boiler room flooded. After temporary repairs in Gibraltar returned to UK March 1941 and thence to repair and refit in Philadelphia April-September 1941. Joined Home Fleet until February 1942 when she went out to the Eastern Fleet. Remained in the Indian Ocean until she returned to UK late 1943. She subsequently became a training ship in May 1944 and was finally sold in 1948.

Ramillies Joined Grand Fleet September 1917. From 1920 to 1935 was in the Mediterranean continuously except for refits. Home Fleet from 1936. After convoy escort duties and a short detachment to Aden, joined the Mediterranean Fleet mid-1940. Bombardment of Bardia 17 August 1940 and Battle of Cape Teulada 27 November 1940. Returned to Atlantic convoy escort early 1941 and on 8 February 1941 her presence with a convoy from Halifax 8 February 1941 prevented attack by *Scharnhorst* and *Gneisenau*. During the chase of the *Bismarck* in May 1941 she was released from a convoy to operate west of the *Bismarck's* position, but did not see any action. Joined Eastern Fleet late 1941. Whilst at anchor in Diego Suarez 30 May 1942 was torpedoed by a Japanese midget submarine and badly damaged. After repairs in Durban and UK she rejoined the Eastern Fleet in September 1943, only to return to UK January 1944. She was part of the bombardment force for the Normandy landings and was the only battleship supporting the landings in the South of France in August 1944. At the end of the war she became an accommodation ship and was sold in 1948.

Above:
***Revenge* in 1942.** *IWM*

Below:
***Royal Sovereign* probably after her refit in Philadelphia in 1943. Radar has been fitted to the director and anti-aircraft guns mounted on B and X turrets.** *IWM*

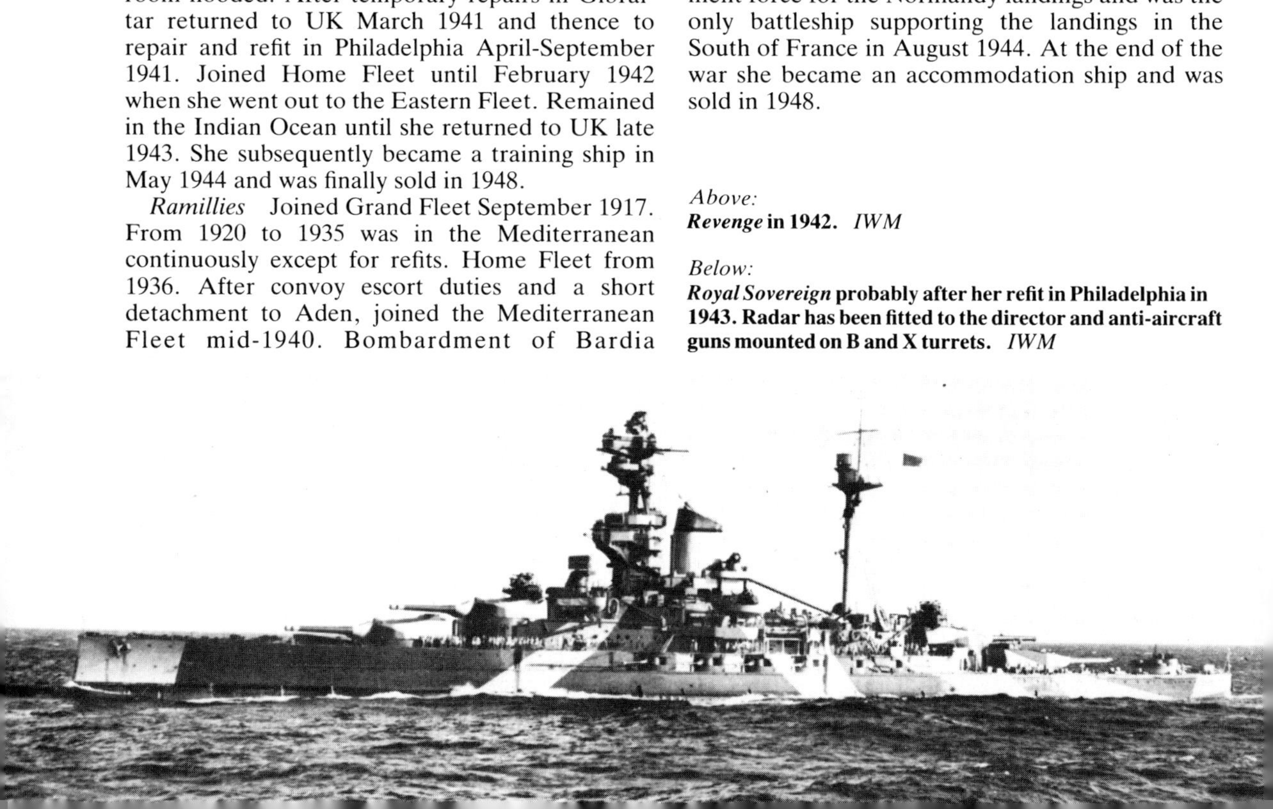

Renown and *Repulse*

AS BUILT
Displacement: 26,500 (26,925)
Length: 794 (244)
Beam: 90 (27.5)
Draught: 26 (8.0)
Armour: 6in (15.2cm) belt, 11in (28.0cm) turrets
Armament: 6×15in; 17×4in; 2×3in; 2×21in torpedo tubes
Machinery: 4 screw, steam turbine
Fuel: Oil
Speed: 32kt
Complement: 967

Unit	Laid Down	Completed	Builder
Repulse	Jan 1915	Aug 1916	John Brown
Renown	Jan 1915	Sep 1916	Fairfield

Design Notes
These ships were the product of Admiral 'Jackie' Fisher's support for the concept of battlecruisers. The value of fast ships to chase the enemy had been demonstrated at the Falklands and Heligoland Bight, but the vulnerability of battlecruisers when operating with the Fleet did not become apparent until after these ships had been launched: Jutland highlighted the weakness of the battlecruisers.

Fisher wanted very fast ships which necessitated long ships, and he also wanted ships that could operate in the Baltic in support of his ideas for an amphibious landing on the north coast of Germany. Therefore he called for ships of limited draught.

Their main armament was limited to six guns and the secondary armament was reduced to the 4in gun — hardly an effective weapon against the destroyers of the time — 15 of which were mounted in triple turrets. Armoured protection was limited, particularly with respect to side armour, but they were the first ships to have bulges designed as an integral part of the hull. The deck armour was improved and anti-flash arrangements were built into the magazines after Jutland. They had the most powerful propulsion system yet fitted into a warship, requiring 42 boilers in six boiler rooms.

Like the 'Queen Elizabeths' and the 'Royal Sovereigns', these ships underwent a series of modifications between the wars.

MODIFICATIONS
1917-24 Flying off platforms were fitted in both ships on B and X turrets. Bulges were modified, and improved side and deck armour was fitted. In *Repulse* eight above-water torpedo tubes were added. In *Renown* 2×3in and 2×4in low angle guns were replaced by 4×4in HA.
1930-41 Both ships underwent reconstruction but *Renown* was reconstructed to a far greater extent than *Repulse*. Therefore at the beginning of World War 2 they were very different ships.

Repulse was reconstructed 1933-36. Deck armour was further improved, and flying off platforms were removed and a catapult fitted. The 4in armament was re-arranged, and 2pdr pom-poms and 0.5in AA guns were added. The submerged torpedo tubes were removed, and there were some changes to the bridge and superstructure.

Renown's reconstruction in 1936-39 was very similar to that carried out in *Warspite*. The 15in guns were modified to a maximum elevation of 30deg. The 4in guns were removed and replaced with 4.5in, and new anti-aircraft armament was added. The submerged torpedo tubes were removed. Deck armour was increased. Completely new machinery and boilers were fitted and the bridge and superstructure were completely rebuilt in the same form as in *Warspite*.
World War 2 During the war both ships had additional anti-aircraft guns added and radar fitted.

THE CLASS IN WORLD WAR 2
The outward appearance of the two ships was very different, *Renown* having the completely rebuilt

Below:
***Renown* in 1943 before her major modification.** *MPL*

bridge and superstructure. The main differences in 1941 from their 'as built' specifications were:

Displacement: Increased by about 5,000 tons
Draught: Increased by about 12in (30cm)
Armour: *Repulse:* 5¾in (14.5cm) deck, 9in (22.9cm) belt
Renown: 5in (12.7cm) deck, 9in (22.9cm) belt

The arrangement of the armour was different in the two ships.

Armament: *Repulse:* 6×15in (20deg elevation); 18×4in; 24×2pdr pom-pom; 16×0.5in AA; 8×20mm
Renown: 6×15in (30deg elevation); 20×4.5in; 24×2pdr pom-pom; 16×0.5in AA
Machinery: *Repulse* unchanged: *Renown* re-engined and re-boilered
Speed: *Repulse* 29kt; *Renown* 30.75kt

Historic Notes

Repulse Joined Grand Fleet August 1916. In action against German light cruisers 17 November 1917 in Heligoland Bight. Refitted 1919-22. In 1923 took part in a world cruise and in 1925 took the Prince of Wales on a foreign tour; otherwise served with the Atlantic Fleet 1922-32. Refit 1933-36 after which she joined the Mediterranean Fleet until 1938, thence returned to UK. Served with the Home Fleet early in the war mainly as a convoy escort. Took part in the Norwegian campaign. On 22 May 1941 sailed with *King George V* and *Victorious* during the hunt for *Bismarck* but had to be detached on 25 May whilst southwest of Cape Farewell to proceed to Newfoundland for fuel. Was in Cape Town in October 1941, having escorted a convoy there, when she was ordered to join the *Prince of Wales* in Ceylon and then proceed to Singapore to counter the threat of Japanese attack. On 8 December 1941 both ships, with an escort of destroyers, sailed from Singapore to attack a Japanese invasion force approaching the Malayan east coast. On 10 December they were attacked by waves of high level bombers and torpedo-bombers. *Repulse* was hit by a succession of torpedoes which caused extensive damage and flooding. She sank after the majority of the crew had been rescued by destroyers.

Renown Joined Grand Fleet September 1916. After the war took the Prince of Wales on foreign tours twice during period 1920-22. Refit 1923-26 and thence joined Atlantic Fleet. Took Duke of York to Australia 1927. Otherwise remained with Atlantic Fleet until 1936. Major reconstruction 1936-39, joined Home Fleet 1939. In action during Norwegian campaign: on 9 April 1940 she engaged *Gneisenau* and *Scharnhorst* west of the Lofotens, obtaining three hits on *Gneisenau* whilst the German ships attempted to escape into a northeasterly gale which limited *Renown's* speed. *Renown* sustained minor damage from three hits. In August 1940 joined Force H at Gibraltar as Flagship. Took part in a number of operations to support convoy movements. In action against Italian battleships and cruisers at Battle of Cape Teulada (Spartivento) on 27 November 1940. Bombarded Genoa 9 February 1941. Sailed with *Ark Royal* and *Sheffield* from Gibraltar in hunt for *Bismarck* May 1941 but was not directly involved in any of the actions. Between August 1941 and February 1943 operated both in the Mediterranean and with the Home Fleet, including covering the Torch landings in North Africa November 1942. Refitted March-July 1943. In late 1943 conveyed the Prime Minister twice, firstly across the Atlantic from Halifax and secondly from UK to Alexandria. 27 December 1943 joined Eastern Fleet and subsequently took part in operations against Sabang, Surabaya, Port Blair and Car Nicobar. Returned to UK April to be reduced to a training role. Sold 1948.

Below:
***Repulse* as modified after her 1933-36 refit.** *Real Photos*

Below:
***Renown* after major modification 1936-39, with a completely new bridge structure.** *IWM*

Courageous and *Glorious*

AS BUILT
Displacement: 18,600 (18,900)
Length: 786 (241)
Beam: 81 (24.7)
Draught: 23 (7.0)
Armour: 3in (7.6cm) belt, 13in (33.0cm) turrets
Armament: 4×15in; 18×4in; 2×3in HA; 2×21in torpedo tubes
Machinery: 4 screw, steam turbine
Fuel: Oil
Speed: 32kt
Complement: 840

Unit	Laid Down	Completed	Builder
Courageous	Mar 1915	Jan 1917	Elswick
Glorious	May 1915	Jan 1917	Harland & Wolff

Design Notes
These ships could be classified simply as large cruisers but they were officially listed as battle-cruisers, although of considerably less displacement than contemporary battlecruisers. They were another product of Admiral Fisher's ideas for amphibious operations on the north coast of Germany in the Baltic. He wanted light, fast ships, heavily gunned. Their armour was very light and thus they were not suitable for operations with the Grand Fleet, although that is where they were stationed after build. Both ships were converted to aircraft carriers after World War 1 and thus pass beyond the scope of this book.

A third ship of the class, *Furious*, was commenced but she was converted to an aircraft carrier whilst she was being built. A flight deck was added covering the forecastle, the forward turret having been done away with.

Historic Notes
Courageous Joined Grand Fleet January 1917. In action with German cruisers in Heligoland Bight in company with *Repulse* and *Glorious* 27 November 1917. Became a gunnery training ship and then spent a period in Reserve after the war. Taken in hand June 1924 for conversion to aircraft carrier.

Glorious Joined Grand Fleet January 1917. In action with German cruisers in Heligoland Bight in company with *Repulse* and *Courageous* 27 November 1917. Like *Courageous* became a gunnery ship after the war and spent a period in Reserve. Conversion to aircraft carrier commenced February 1924.

Above:
***Glorious*. One of the triple 4in mountings can be seen amidships, a mounting which was not a great success. The triple mountings had first been fitted in *Repulse* and *Renown*.** *IWM*

Below:
***Courageous* during World War 1.** *IWM*

VIII
Post-World War 1 Construction 1920-60

The battleship programme after World War 1 was dominated by the provisions of the Washington Treaty, agreed in 1922. This was effectively a disarmament agreement, the main effects on Britain being:

- A reduction of the capital ship fleet to 21 operational ships.
- Parity with the USA, measured in total tonnage, Japan, France and Italy all being allowed lower tonnages.
- Britain allowed to build two new ships (*Nelson* and *Rodney*).
- Ships to be limited to 35,000 tons and 16in gun maximum.
- Old ships could only be replaced when 20 years old.

In the 1920s only *Hood*, *Nelson* and *Rodney* were completed and it was not until 1936 that a new construction programme was commenced, the first of the 'King George V' class ships being laid down in early 1937. Thus the reconstruction programme carried out in the 1930s on some of the 'Queen Elizabeth' class and on *Renown* was vital in providing ships which were fit to fight in World War 2.

The emphasis in design in this period was on better armoured protection, in particular against torpedoes and also in the provision of armoured decks able to withstand long-range plunging fire. Defence against aircraft attack assumed ever-increasing importance and all the ships built in this period had considerable increases in their close-range anti-aircraft armament as time went on. The improvements in main armament gunnery were gradual, the only major advance, although a significant one, being the introduction of radar for gunnery control. This allowed not only accurate ranging but also blind fire against an unseen target.

Propulsion machinery continued to improve with small-tube, higher pressure boilers allowing greater pressures to be developed. The speed of the later classes of battleship equalled that of the earlier battlecruisers.

Hood

AS BUILT
Displacement: 41,200 (41,860)
Length: 860 (260)
Beam: 104 (31.8)
Draught: 28½ (8.8)
Armour: 12in (30.5cm) belt, 15in (38.1cm) turrets, 1½in (3.8cm) decks
Armament: 8×15in; 12×5.5in; 4×4in AA; 6×21in torpedo tubes
Machinery: 4 screw, steam turbine
Fuel: Oil
Speed: 32kt
Complement: 1,470

Unit	Laid Down	Completed	Builder
Hood	Sep 1916	Mar 1920	Clydebank

Design Notes
Hood is usually considered to be a battlecruiser. However, by the time she was under construction, the lessons concerning the vulnerability of the lightly-armoured battlecruisers had been learned and the protection incorporated in the design of *Hood* was increased considerably. The requirement remained for a fast ship which demanded a long hull. Therefore, in common with earlier battlecruisers, she was considerably longer and heavier than the contemporary battleships, the more so because her design incorporated the armoured protection of a battleship. She was the first large warship to have small-tube boilers which considerably increased the power of her propulsion and which gave her a maximum speed of 32kt.

Her armament was conventional although her 15in mountings were modified to allow a maximum elevation of 30deg. Her secondary battery was designed as protection against destroyers and represented a return to an open battery positioned comparatively high amidships, the lower enclosed battery of the 'Queen Elizabeth' class having proved unsatisfactory in a seaway. At the time of building, her anti-aircraft armament was minimal.

Top:
***Hood* on sea trials, steaming at maximum speed.** *IWM*

Above:
***Hood* at the Coronation Review, 20 May 1937.** *IWM*

MODIFICATIONS

Like the other classes between the wars, *Hood* underwent some modifications although no major reconstruction work was undertaken. By the outbreak of World War 2 she still had the armour and machinery with which she was built. Reconstruction had been proposed to be undertaken from 1939 or shortly thereafter but the outbreak of war prevented her being taken in hand. In a series of refits between 1929 and 1939, the following major changes were made:

- Pom-pom and 0.5in machine gun anti-aircraft mountings fitted.
- 4in HA guns mounted.
- Aircraft catapult fitted.

In early 1940 all the 5.5in guns were removed and additional anti-aircraft armament mounted, including the 20-barrelled rocket launchers known as UPs (Unrotated Projectile). Gunnery radar was also fitted. Therefore by mid-1941 her armament consisted of: 8×15in; 14×4in dual-purpose in twin mountings; 24×2pdr pom-poms; 16×0.5in MGs in quadruple mounts; and 5×20-barrelled UP launchers.

Historic Notes

In 1920 *Hood* joined the Fleet as Flagship of the Battle Cruiser Squadron. In 1923 she went on a world cruise with *Repulse* to South Africa, the Far East, Australia, across the Pacific to the United States and thence home, taking 10 months in all. She remained with the Home and Atlantic Fleets until 1936, undergoing a long refit 1929-31, and a number of shorter refits. In 1936 she joined the Mediterranean Fleet. After a short refit in early 1939 she rejoined the Home Fleet. In June 1940 was part of Force H formed at Gibraltar and subsequently took part in the action against the French Fleet at Oran in early July 1940. Rejoined the Home Fleet August 1940. On 19 May 1941 *Hood* in company with *Prince of Wales* sailed from Scapa Flow to join the hunt for the *Bismarck*, and in the early morning of 24 May both ships engaged the *Bismarck* and the accompanying cruiser *Prince Eugen*. *Bismarck's* fifth salvo straddled *Hood* and one, possibly two, hits were obtained which caused a massive explosion in the vicinity of the main mast, believed to be a magazine blowing up. This explosion broke the ship in two and she sank rapidly with the loss of all but three of her ship's company.

Nelson and *Rodney*

AS BUILT
Displacement: 33,950 (34,490)
Length: 710 (217)
Beam: 106 (32.4)
Draught: 30 (9.2)
Armour: 14in (35.6cm) belt, 16in (40.7cm) turrets, 6in (15.3cm) decks
Armament: 9×16in; 12×6in; 6×4.7in AA; 8×2pdr pom-poms; 2×24½in torpedo tubes
Machinery: 2 shafts, steam turbine
Fuel: Oil
Speed: 23kt
Complement: 1,300

Unit	Laid Down	Completed	Builder
Nelson	Dec 1922	Jun 1927	Armstrongs
Rodney	Dec 1922	Aug 1927	Cammell Laird

Design Notes
These two ships had to be designed within the constraints of the Washington Treaty which allowed 16in gun ships to be built to match the American 'Maryland' and Japanese 'Nagato' classes but which limited the displacement to a maximum of 35,000 tons. The unusual layout of the ships with the three turrets forward of the bridge structure was largely so that the area requiring the maximum armoured protection could be concentrated as much as possible. In order that the armoured protection as a whole could reach adequate standards, the whole emphasis of the design was on saving weight elsewhere by such measures as using lighter steels for unarmoured sections and even down to the use of aluminium fittings for furniture, etc.

The main machinery layout also was unusual. The design reverted to two shafts and the engine rooms were forward of the boiler rooms, which allowed the funnel to be placed in a conventional position abaft the bridge and also allowed a better arrangement of machinery within the hull.

Overall the armour was superior to earlier ships, in particular with respect to deck armour which was much increased. The 16in guns were a new design as were the triple turrets, and the early days

Below:
***Nelson* as completed. The new triple 16in mountings were very unreliable in their early years.**
Vickers Shipbuilding & Engineering Ltd

Above:
***Rodney*, probably in 1942. Radar was fitted during her refit in Boston in 1941.** *IWM*

of this armament were fraught with many technical problems. The arcs of fire were very restricted not only by the fact that all the turrets were forward but also because it was found that fire abaft the beam caused considerable damage to the bridge, particularly if X turret was firing. They were fitted with an improved anti-aircraft armament, being the first ships to be fitted with 2pdr AA guns for close-range protection.

They were regarded as somewhat ugly ships and had a reputation for being difficult to manoeuvre. Nevertheless they had an appearance of power, enhanced by the tower bridge structure which was to become the standard design for future battleship structures and was incorporated in the ships which were reconstructed between the wars.

MODIFICATIONS

In common with their contemporaries, the main modifications were to anti-aircraft armament. Additions were made before 1939 and further guns were added during the war. Apart from the fitting of an aircraft catapult in the 1930s and the fitting of radar during the war, no other major work was undertaken except some attempts to strengthen the forward part of *Rodney's* hull which began to show signs of weakness. By the end of World War 2 their armament was: 9×16in; 12×6in; 6×4.7in AA; 48 (*Nelson*)/44 (*Rodney*)×2pdr pom-poms; 16×40mm Bofors (*Nelson* only); and 61/68×20mm.

Historic Notes

Nelson All her prewar service was with the Atlantic and Home Fleets. On 4 December 1939 she was damaged by a magnetic mine off Loch Ewe, being repaired in Portsmouth during the next eight months. Thence mainly with the Home Fleet until September 1941. On 27 September she was hit by a torpedo from an aircraft whilst escorting a convoy from Gibraltar to Malta, and had to return to UK for repairs which lasted until April 1942. Joined the Mediterranean Fleet escorting a number of Malta convoys and took part in bombardments supporting the landings in North Africa, Sicily and at Salerno. She returned to UK to take part in the Normandy landings but on 18 June 1944 was extensively damaged by two magnetic mines in the English Channel. Repairs lasting six months were carried out in Philadelphia after which she proceeded to the Indian Ocean and took part in the final operations against the Japanese in Malaya. She returned to the Home Fleet at the end of 1945 but became non-operational in late 1946 and was finally sold in 1948.

Rodney Served with the Atlantic Fleet 1928-32 and thereafter with the Home Fleet. During operations off Norway in April 1940 she was hit by an armour-piercing bomb which caused minor damage. Employed as a convoy escort for a period. In May 1941 she detached from a convoy to join *King George V* and took part in the final action against the *Bismarck* on 27 May which led to the destruction of the German ship after numerous hits from both British battleships. Served with Force H for a period in late 1941. Refit December 1941-May 1942 and thence supported landings in North Africa, Sicily and at Salerno. Took part in numerous bombardment operations off Normandy June-September 1944 and, after a final operation with a Russian convoy in late 1944, finished the war in Scapa Flow. Sold 1948.

'King George V' Class

AS BUILT
Displacement: 36,750 (37,335)
Length: 745 (229)
Beam: 103 (31.5)
Draught: 35½ (10.8)
Armour: 15in (38.1cm) belt, 16in (40.7cm) turrets, 6in (15.3cm) deck
Armament: 10×14in; 16×5.25in; 32-48×2pdr pom-poms; 0, 6 or 18×20mm Oerlikon
Machinery: 4 shaft, steam turbine
Fuel: Oil
Radius: 15,000nm at 10kt
Speed: 28kt
Complement: 1,500

Unit	Laid Down	Completed	Builder
King George V	Jan 1937	Dec 1940	Vickers
Prince of Wales	Jan 1937	Mar 1941	Cammell Laird
Duke of York	May 1937	Nov 1941	Clydebank
Anson	July 1937	Jun 1942	Swan Hunter
Howe	Jun 1937	Aug 1942	Fairfield

Design Notes
These ships were very much a result of the termination of the Washington Treaty, and the building programme was initiated largely to match the programmes of Germany, Italy and indeed France. Their design was a reversion to the 14in gun but mounted in the unique arrangement of quadruple turrets forward and aft with a twin B turret. The 5.25in guns in twin turrets were designed for both a surface and anti-aircraft role. The anti-aircraft close-range armament varied from ship to ship, the later ships having more such armament incorporated at build.

The overall layout of the ships represented a return to 'conventional' ideas after the somewhat unusual layout of the *Nelson* and *Rodney*. The armoured protection was comprehensive, with very thick deck armour protecting the vital areas, particularly the magazines. The anti-torpedo bulges of earlier classes were dispensed with: instead the wing tanks below the armoured belt were designed to absorb the impact and explosion of a torpedo hit.

MODIFICATIONS
The significant modifications carried out to the ships during their short lives were the fitting of, and then improvement to, radar, and a steady increase in the amount of close-range anti-aircraft armament fitted. By the end of World War 2 each ship had a different mix of anti-aircraft armament, although the typical fit was of the order of: 88×2pdr pom-poms; 8×40mm Bofors; and 55×20mm Oerlikons.

Historic Notes
King George V Joined the Home Fleet on commissioning. In May 1941 was the Flagship of the Home Fleet and sailed on 22 May from Scapa Flow with the C-in-C, Admiral Tovey, embarked to join the hunt for the *Bismarck*. Took part in the final action against *Bismarck* with *Rodney* 27 May 1941. After further service with the Home Fleet which included a number of operations covering convoys, she supported the Sicily landings in July 1943, and took part in the occupation of Taranto September 1943. After a refit in early 1944, she joined the Far East Fleet and took part in the final

Below:
***Prince of Wales* in 1941. Gunnery radar was fitted in mid-1941 and can be seen above the main director.** *IWM*

Above:
Anson about 1942 with an improved radar fit. *IWM*

Left:
Anson shortly after World War 2. The armoured belt along the waterline amidships can be clearly seen. *Real Photos*

operations against Japan. On 29 July 1945 she was the last British battleship to fire her guns in wartime when she bombarded targets south of Tokyo. After the war she served in the Home Fleet before going into Reserve and finally being scrapped in 1957.

Prince of Wales Whilst still not fully worked up she joined in the hunt for the *Bismarck* in May 1941. On 24 May she engaged the *Bismarck* along with *Hood*, during which action *Hood* was sunk. Although *Prince of Wales* achieved three 14in hits on the *Bismarck*, she herself was hit a number of times and suffered teething troubles with her main armament which eventually forced her to break off the engagement. In July 1941 she took the Prime Minister to Placentia Bay, Newfoundland to meet President Roosevelt. Supported convoy operations in the Mediterranean September 1941. In October 1941 she was sent to the Far East via South Africa to join with *Repulse* in countering the Japanese threat. On 8 December 1941 both ships sailed from Singapore to oppose a possible Japanese landing on the east coast of Malaya, and on 10 December both ships were attacked successively by both bomb and torpedo-armed Japanese aircraft. After a number of torpedo and bomb hits, *Prince of Wales* finally sank, albeit after some 1,250 of her ship's company had been rescued by destroyers.

Duke of York Served with the Home Fleet continuously until 1945, covering many convoy operations, particularly those en route for Russia. Whilst covering such a convoy on 26 December 1943 she engaged the *Scharnhorst*, assisted by the accompanying British cruiser force. *Duke of York's* gunfire was particularly accurate, being supported by accurate radar ranging, and eventually slowed *Scharnhorst* down enough for the cruisers and destroyers to finish her off with torpedoes in what has become known as the Battle of the North Cape. In May 1945 she sailed to join the Far East Fleet, being present at the final surrender of the Japanese in Tokyo Bay in August 1945. After the war she served with the Home Fleet for a time before being placed in Reserve and finally being scrapped in 1957.

Anson Joined the Home Fleet in June 1942 and was mainly engaged in supporting Russian convoys although she took part in no significant operations. In March 1945 she was sent to the Far East Fleet and took part in the re-occupation of Singapore. Remained in the Far East until July 1946 when she returned for a short refit before taking up training duties in UK. Scrapped 1957.

Howe Joined the Home Fleet August 1942. Supported the Sicily landings July-September 1943. Joined the Eastern Fleet July 1944 and thereafter took part in a number of bombardment operations against the Japanese. After a refit in the latter part of 1945, became a training ship in home waters before going into Reserve and finally being scrapped 1957.

Vanguard

Right:
***Vanguard*, the end of an era.** *IWM*

Displacement: 44,500 (45,210)
Length: 814 (250)
Beam: 108 (33.0)
Draught: 36 (11.0)
Armour: 14in (36.6cm) belt, 13in (33.0cm) turrets, 6in (15.3cm) deck
Armament: 8×15in; 16×5.25in; 73×40mm Bofors
Machinery: 4 shaft, steam turbine
Fuel: Oil
Speed: 30kt
Complement: 2,000

Unit	Laid Down	Completed	Builder
Vanguard	Oct 1941	Apr 1946	Clydebank

Design Notes

Vanguard was the result of an emergency war building programme. In order to hasten her completion, the main armament was provided from the guns and turrets removed from *Courageous* and *Glorious* during their conversion to aircraft carriers. When fitted in *Vanguard*, the turrets were modified to improve the elevation of the guns and thicker turret armour was provided. The overall design of the ship owed a lot to the 'King George V' class. In particular the armoured protection was very similar.

During construction the lessons of World War 2 were becoming apparent. Changes were made to some of the damage control arrangements and the original outfit of close-range anti-aircraft armament was increased considerably. The very flat forecastles of the 'King George V' class had made them very wet in any seaway so *Vanguard* was given a pronounced sheer. With her raked cowls on the funnels to keep gases clear of the bridge, her overall appearance was very handsome, and she had a reputation as a very good seaboat.

1 Paint store
2 Watertight compartment
3 Ratings' heads and bathrooms
4 Naval stores
5 Soap and tobacco store
6 Inflammable store
7 Arctic clothing store
8 Capstan machinery
9 Library
10 Reading room
11 School
12 Anchor gear store
13 Cable locker flat
14 Cells
15 Asdic office
16 Engineers' spare gear
17 Cable locker
18 Officers' store
19 Provision room
20 Pump room
21 Fresh water tanks
22 Boys' mess
23 Telegraphists' mess
24 Clothing store
25 Provision store
26 Painted canvas room
27 Store
28 Seamen's mess
29 Petty Officers' mess
30 Flour store
31 Cold and cool rooms
32 Gunner's store
33 Cordite handling room
34 Shell handling room
35 Shell room
36 Magazine
37 Store
38 Canteen
39 Dining hall
40 Servery
41 Sickbay
42 Bathroom
43 Operations room
44 Transmitting station
45 Gyro compass room
46 5.25in shell room
47 Bofors magazine
48 5.25in magazine
49 5.25in casemate
50 Drying room
51 Workshop
52 Bathroom
53 Soot removal plant
54 Boiler uptake space
55 Boiler room
56 Cafeteria space
57 Stokers' mess
58 Computer and breaker rooms
59 Engine room
60 Marines' mess
61 Messdeck

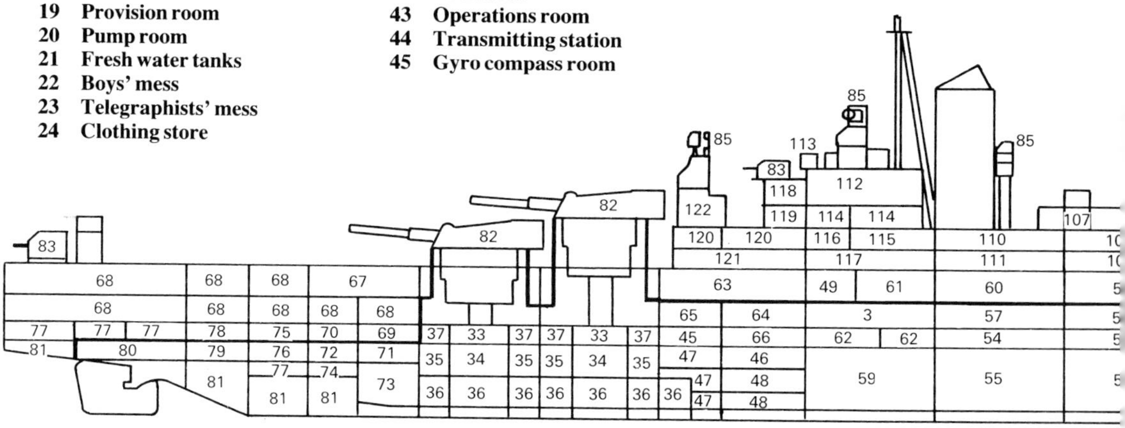

Historic Notes

Her career was short and spent mainly with the Home Fleet. She took King George VI on a royal tour to South Africa in 1947. After 1949 she was used mainly in a training role and was placed in Reserve in 1954. As the last British battleship she was towed out of Portsmouth on her way to the breakers in 1960, ending an era.

HMS *Vanguard*

62 Workshops
63 Cinema
64 Naval store
65 Medical distribution centre
66 Breaker room
67 Officers' bathroom
68 Cabins
69 Church
70 Printing office
71 Pump room
72 Store
73 Gland space
74 Spirit room
75 Officers' store
76 Wardroom store
77 Store
78 Capstan machinery
79 Awning store
80 Steering gear compartment
81 Watertight compartment
82 Twin 15in mounting
83 Bofors mounting
84 Compass platform
85 Gunnery director
86 Charthouse
87 Captain's quarters
88 Admiral's bridge
89 Cabins
90 Store
91 Galley
92 Wireless office
93 Main signal office
94 Cryptographic office
95 Downtake
96 Lobby
97 Signal house
98 Y office
99 Heads
100 Office
101 Store
102 Office
111 Petty Officers' mess
112 Ready use magazine and shelter
113 Close-range gunnery director
114 Cabins
115 Chief of Staff's cabin
116 Pantry
117 Wardroom
118 Predictor room
119 Wardroom galley
120 Admiral's quarters
121 Wardroom anteroom
122 After conning position
103 Meteorological office
104 Radar office
105 Workshop
106 Workshop
107 Engine room exhaust
108 CPO and POs' mess
109 Galley and bakery
110 Gunroom

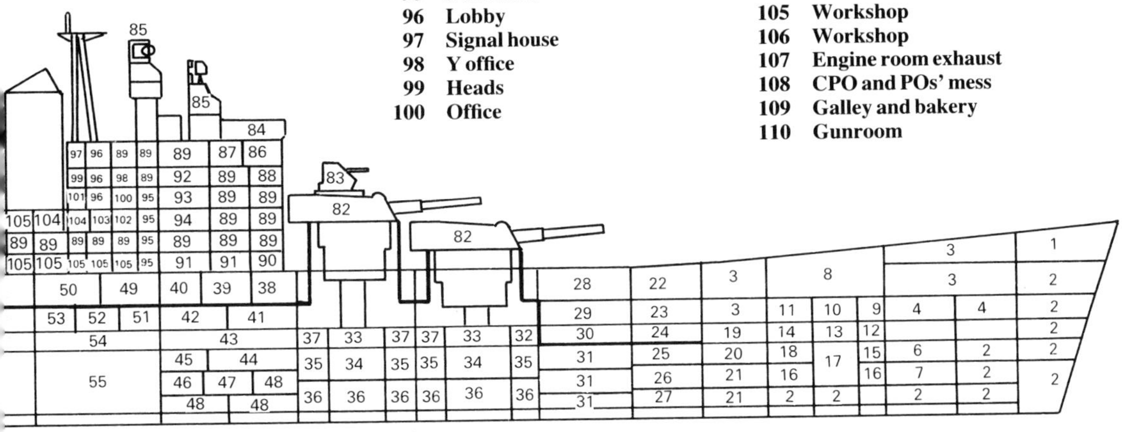

Figure 4 — HMS *Vanguard* internal layout

Appendices
1
Facts and Figures

Virtually any statistic one takes from the history of battleships shows a dramatic increase over the 100 years of the battleship era. The following graphs show just some of the significant changes which occurred.

Figure 5 Displacement was fairly stable, on average, until 1880 and then rose steadily thereafter. The greater displacement of the later battlecruisers is evident.

Figure 6 Complement varied considerably in the early years due to variations in ship size, gun armament and sailing rig. The effect of the abolition of sail around 1890 is apparent. The large increases from World War 1 onwards are mainly because of the increase in the number of guns fitted, particularly as more and more anti-aircraft guns were fitted.

Figure 7 Speed showed a steady increase over the years. *Dreadnought* and the 'Queen Elizabeth' class show up as being particularly fast for their times and the dramatic speed increases achieved for the battlecruisers can be clearly seen.

Right:
***Howe* at speed.** *IWM*

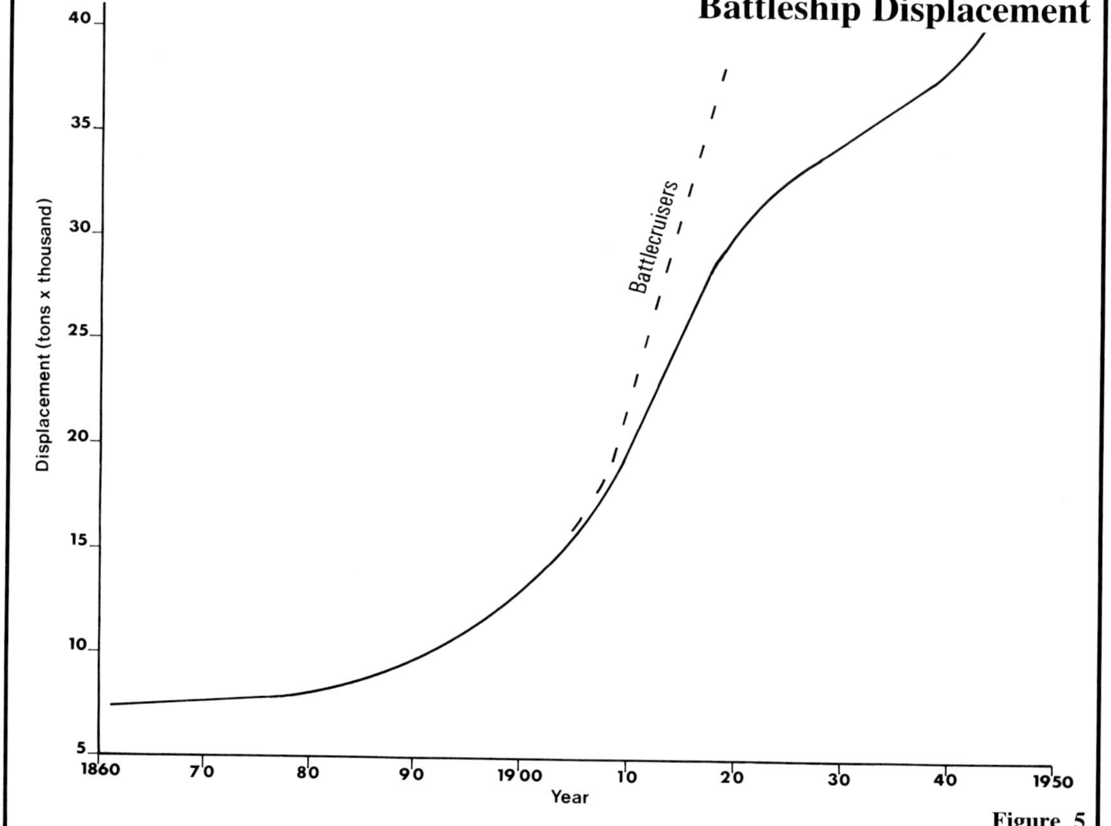

Figure 5

Figure 8 Up to 1900 gunnery fighting ranges were considered to be of the order of one mile or less. The breakthrough achieved in the first 10 years of the 20th century pushed the ranges out to six miles or more. Fighting ranges were, in practice, very similar in both world wars.

Figure 9 The effect of changes in the main armament size is apparent. The large increase around 1911-12 was due to the re-introduction of the 13.5in gun in the 'Orion' class, which mounted 10 such guns. Also the significance of the 16in guns in the 'Nelson' class is clear.

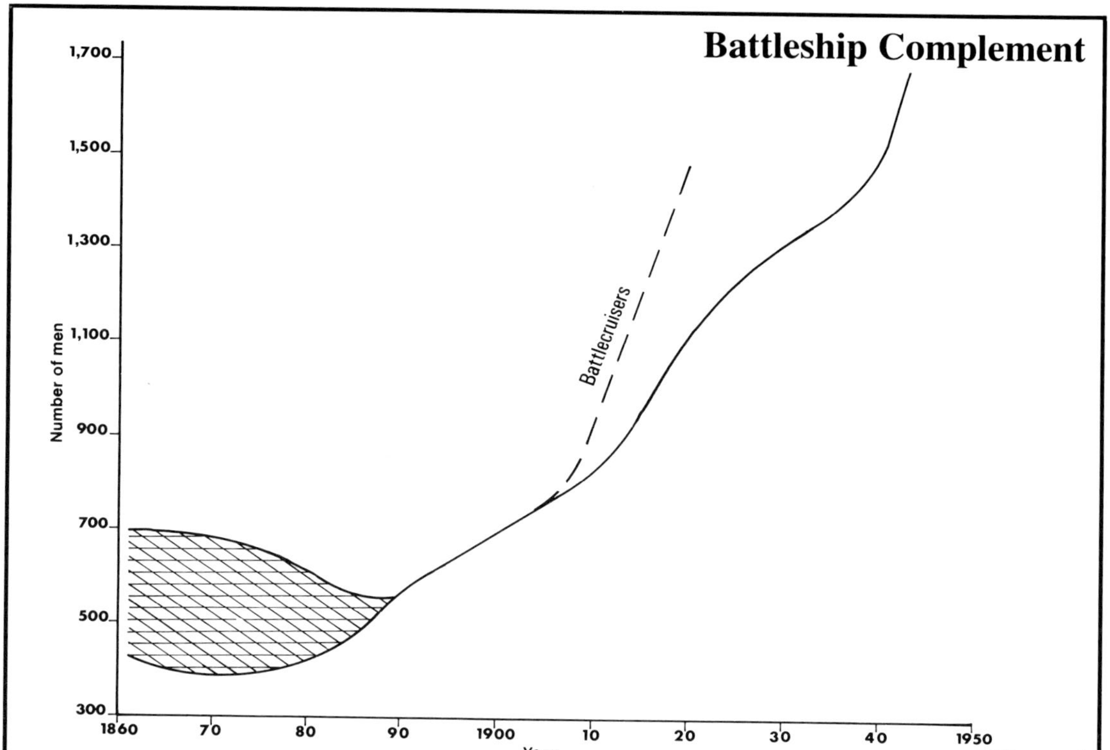

Figure 6

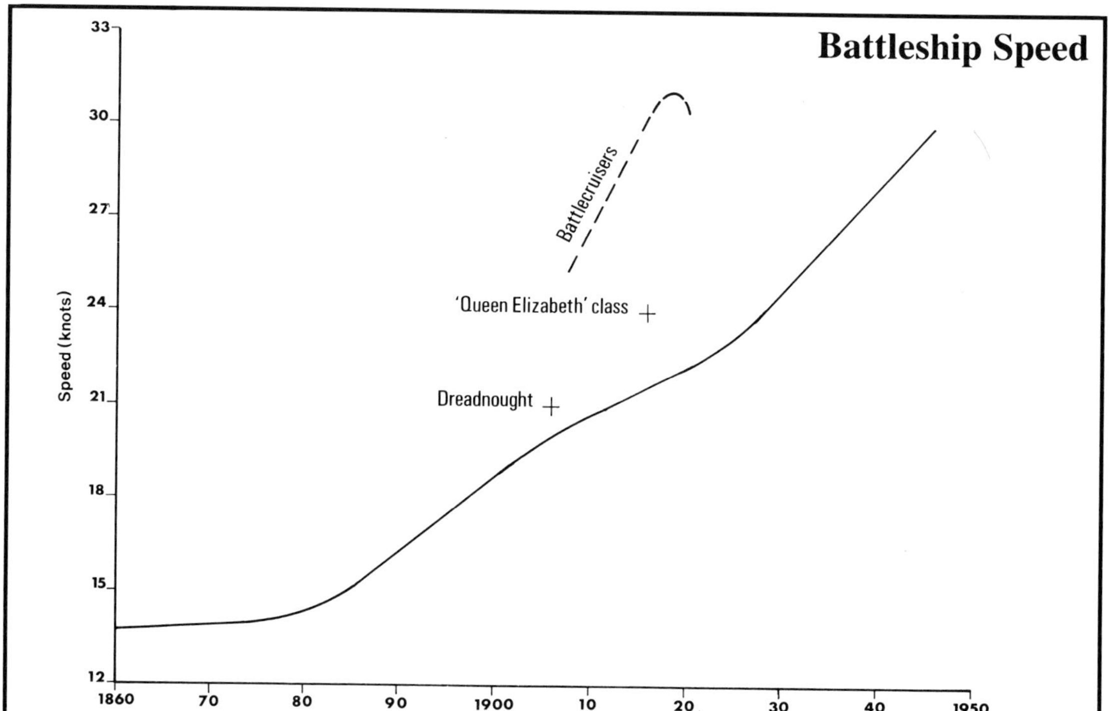

Figure 7

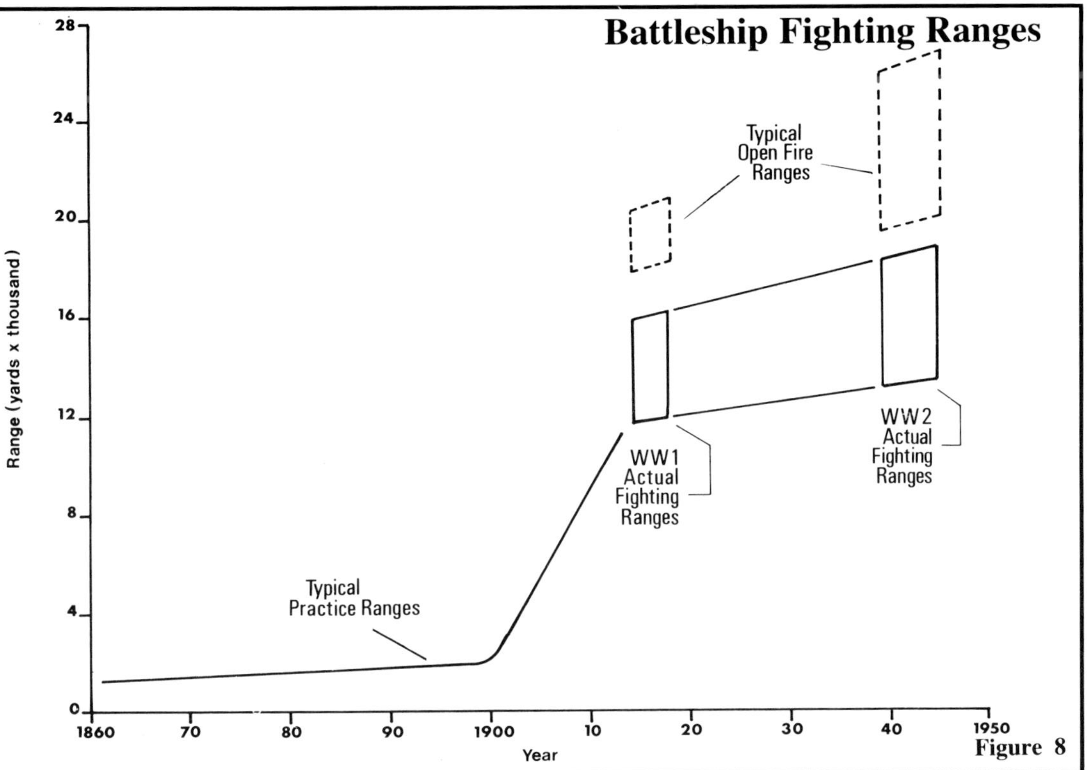

Figure 8

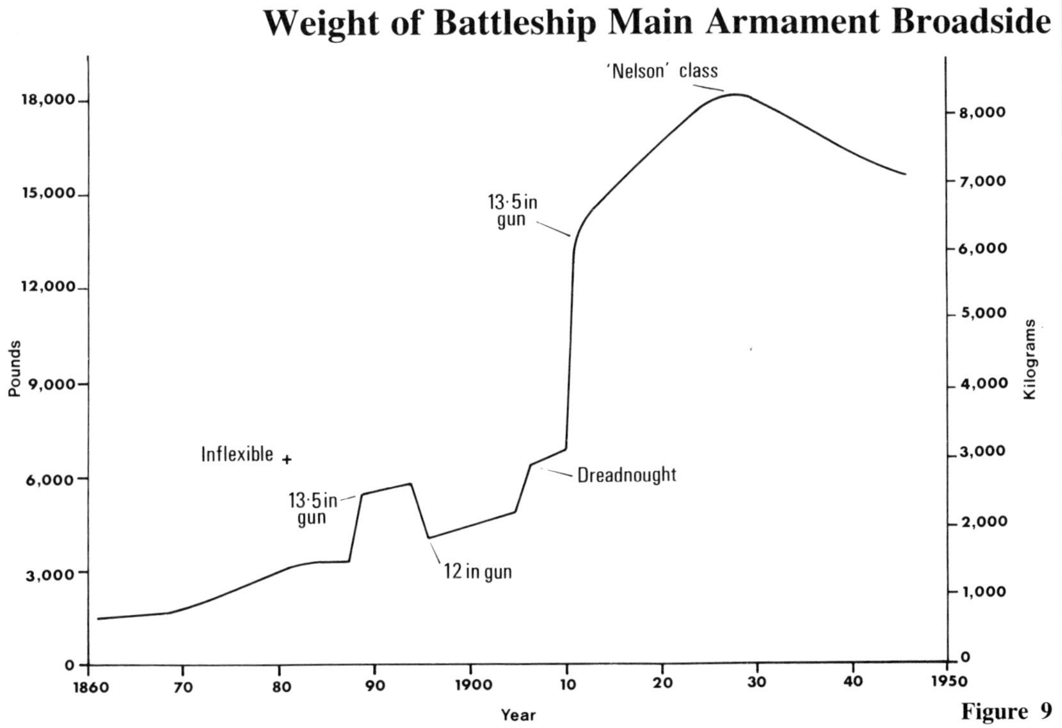

Figure 9

2
Main Armament

Main Armament Layout

Figure 10 shows examples of the various main armament layouts adopted in the various classes.

Warrior (1861) Guns disposed along the length of the ship, a pattern unchanged since *Victory*. The larger 110pdrs were placed centrally in the ship.

Hercules (1868) A smaller number of guns have been placed in a central, strongly-armoured battery.

Monarch (1896) Amongst the first of the turret ships, the two turrets were placed amidships where they could only fire on the beam.

Devastation (1873) The turrets have been placed nearer the extremities of the ship and at least one turret can fire on any bearing.

Inflexible (1881) A return to placing the turrets amidships, as part of a strong, central citadel. The turrets are displaced off the centreline to allow axial fire forward and aft.

Conqueror (1886) An unsatisfactory arrangement, representing the weakest main armament ever mounted in just a single turret, capable of firing over the forward arc only.

'Admirals' (1889) A return to main armament forward and aft, in this class mounted in barbettes.

Canopus (1899) By this stage the turret had returned and the main armament was positioned forward and aft on the centreline.

King Edward VII (1905) Ships of this class had two twin 12in turrets on the centreline plus four 9.2in single turrets, two on each side.

Lord Nelson (1908) This class took the mixed arrangement one stage further by having two twin 12in turrets, four twin 9.2in turrets and two single 9.2in turrets, 14 guns in all.

Dreadnought (1906) Mounted a single size of main armament, disposed in five twin turrets.

Neptune (1911) The first ship to have superimposed turrets, in this case the two after turrets. The two midships turrets were able to fire across the ship, to give a full 10-gun broadside.

King George V (1912) Had superimposed turrets both forward and aft.

Queen Elizabeth (1915) With the fitting of the 15in gun, the number of turrets was reduced to four, superimposed forward and aft.

Nelson (1927) Washington Treaty requirements dictated the unusual arrangement of three triple turrets, all forward, the forward two superimposed.

King George V (1940) Unique in having quadruple turrets, allowing 10 14in guns to be mounted in just three turrets.

Left:
***Royal Oak*, the power of her eight 15in amply demonstrated.**
IWM

Main Armament Layouts

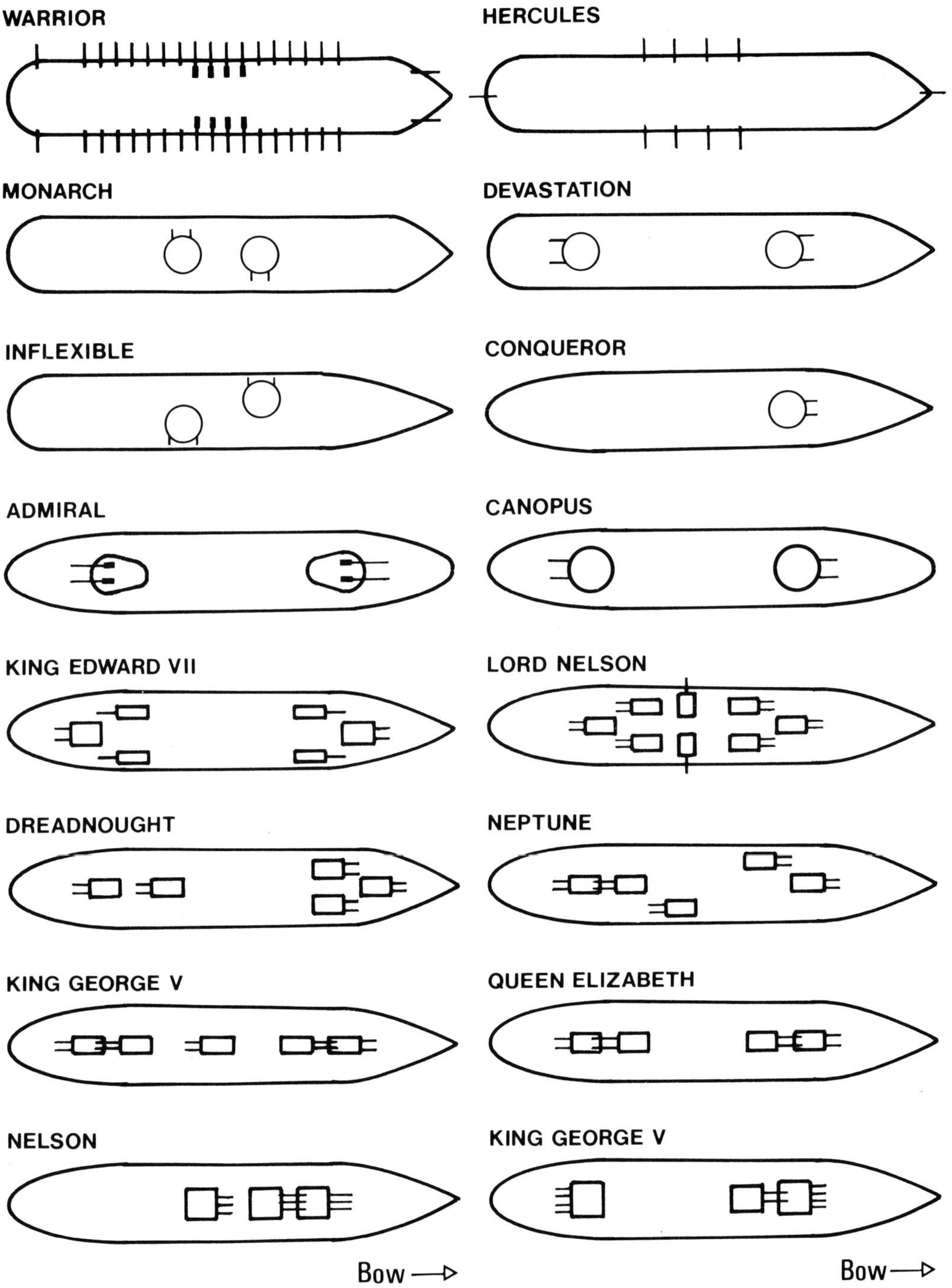

Diagrammatic – not to scale

Figure 10

Gunnery Fire Control

When *Warrior* went to sea, fire control had advanced little since Trafalgar: guns were aimed individually with no central control, and were fired when the ship's roll had brought them to a suitable elevation. Capt Percy Scott introduced the first major changes from about 1870 onwards. He invented a much improved gun carriage which allowed the elevation of the gun to be continuously adjusted as the ship rolled. This 'continuous aim', where the gun captain kept the gun constantly aimed at the target while compensating for roll, improved the accuracy of gunnery.

As ranges started to increase in the mid-1890s, optical range-finders started to come into service. Accurate range-finding was essential if the first few salvoes were to be effective in an engagement.

By the time of World War 1, various mechanical aids to fire control had been introduced and new concepts adopted:

Dreyer table This table was essentially a mechanical computer, taking in the bearing, range, spotting corrections, wind, etc, and generating a bearing and range upon which to fire the guns.

Dumaresq This kept the 'rates', ie it matched the rate at which the target bearing and range were changing and fed this information to the Dreyer table so the target could be tracked continuously.

Director The director was mounted high in the ship and from here the bearing and range of the target were measured.

Control position This was usually the upper fighting top on the foremast, with the director mounted just below it. From here the guns were centrally controlled and spotting corrections were passed to the Dreyer table.

Figure 11 shows how the various elements combined.

By the time of World War 2, the system had been modified somewhat, with the Director and Control Position being combined into a single, larger Director. By the middle of the war, radar range finding was being introduced and by 1944 ships were able to fire entirely on radar information. Figure 12 shows the final arrangement.

Below left:
Air Defence Postion on board a 'King George V' class ship. The aircraft look-outs report sightings to the Air Defence Officer who can be seen at his sight. This sight fulfils the same function as the Captain's Sight shown in the World War 2 fire control diagram — it indicates the target to the director, ie the director will follow where the sight is pointing, allowing the director to be quickly locked on to the target. A pom-pom director can be seen lower down. *IWM*

Below:
A 5.25in gun director on *King George V*. The Director Officer is seated in the middle (bare-headed) with the director layer and trainer either side, whose task it is to follow the target in elevation and bearing respectively. Lower down at the front, the range taker is operating the optical range-finder. Notice how exposed this position is. *IWM*

Right:
The end result — 15in guns engaging the target. The photograph is taken on *Valiant* with *Barham* and *Warspite* ahead. *IWM*

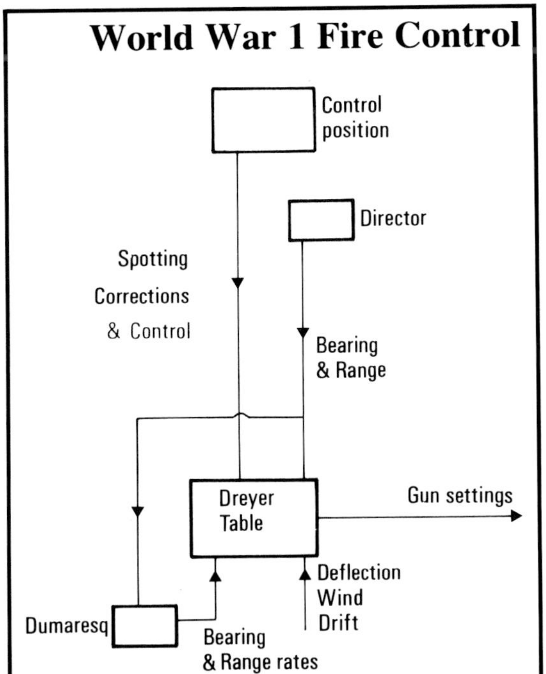

Figure 11

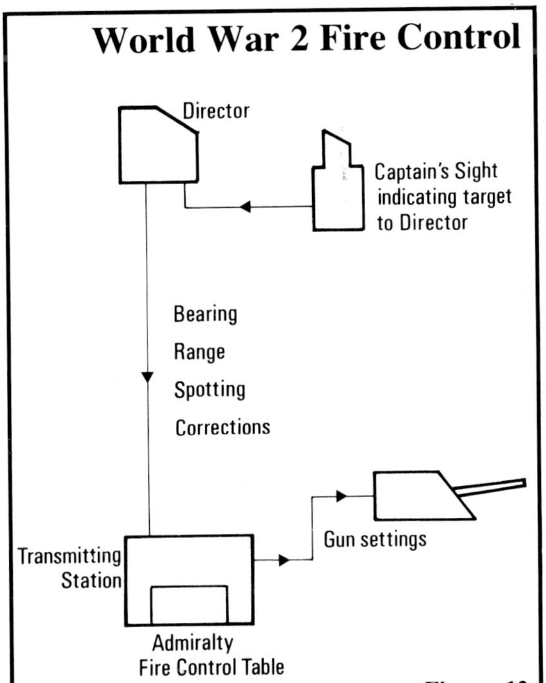

Figure 12

In World War 2 the armament of battleships reached its zenith. A wide variety of guns was fitted. The photographs below show examples of the guns which fought in World War 2 and Figure 13 shows the engagement envelopes for the main guns in use. Note that all the guns except the main armament have an anti-aircraft role, although the 4.5in and 5.25in guns were dual-purpose.

Left:
After quadruple 14in turret on board *King George V*. Two UP launchers are mounted on top of the turret. *IWM*

Far below left:
A 4.7in gun on a 'Nelson' class ship. When originally fitted, these guns had no shields and the gun's crew was completely exposed. *IWM*

Below left:
Inside a triple 16in turret on board a 'Nelson' class ship. *IWM*

Bottom left:
Twin 40mm Bofors mounting. Like the pom-pom, this also fired a 2lb shell. *IWM*

Right:
Twin 5.25in turret on board a 'King George V' class ship. The high elevation of one of the guns demonstrates the dual role of this mounting, having not only a surface capability but also an anti-aircraft one. *IWM*

Below:
Eight-barrelled 2pdr pom-pom mountings on board *King George V*. This was a close-range, anti-aircraft mounting firing a 2lb, 40mm shell. *IWM*

Above:
Two single 20mm Oerlikon mountings. By the end of the war, the 'King George V' class had been fitted with up to 55 of these guns, required in particular for duty in the Far East. *IWM*

Above right:
An Unrotated Projectile mounting as fitted on board 'King George V' class ships. These 20-barrelled mountings fired Parachute and Cable rockets which, on reaching a fixed height, released parachutes which dangled small anti-aircraft bombs on wires beneath them. They were fitted early in the war but were removed as more close-range guns became available. The UP was not a success. *IWM*

Right:
A submerged torpedo tube on board HMS *Hood*. Battleships had been fitted with submerged torpedo tubes since before World War 1, but by 1939 most of the tubes had been removed. *Wright & Logan*

Gun Envelopes

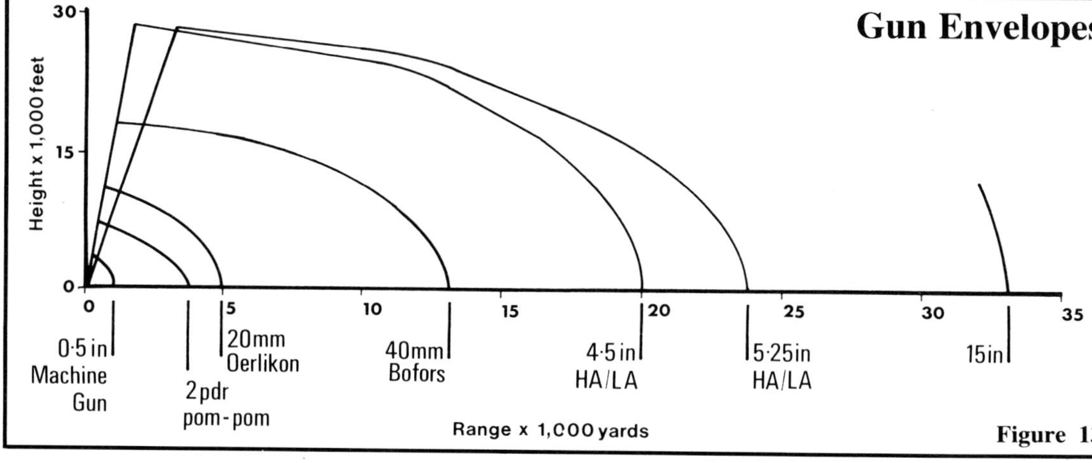

Figure 13

3
Aircraft

It is not generally appreciated that all battleships carried aircraft in the period from about 1918 to about 1944-45. The year of original fit and eventual removal of the aircraft varied from ship to ship, but by the end of World War 1 most ships had been fitted with the first type of launching device, the flying-off platform.

Ships either carried fighter aircraft (Sopwith Pups or Camels) for use against Zeppelins, or spotter-reconnaissance aircraft (Sopwith 1½ Strutters). Flying-off platforms were soon replaced by catapults, the first catapult launch being made by a Fairey IIID in October 1925. As catapults evolved, various fits were tried, including on top of gun turrets, on the quarterdeck and, later, athwartships amidships, with aircraft hangars just forward of the catapult.

Between the wars, more modern aircraft were adapted for shipborne use. The original aircraft flown off the platforms could not be recovered — they either had to fly to land or the pilot had to ditch or bale out. Later, aircraft were fitted with floats so that, in calm conditions at least, the aircraft could land near the ship and be recovered by crane. By the beginning of World War 2 the aircraft in service in battleships were the Fairey Seafox, Supermarine Walrus and the ubiquitous Swordfish. Their role by this stage was almost entirely reconnaissance and gunnery spotting. Although not a battleship action, one aircraft was launched prior to the Battle of the River Plate and spotted the fall of shot throughout the action. In the Second Battle of Narvik, *Warspite* used her Swordfish to good effect, scouting ahead of the ship and thus finding the German destroyers hidden in the side fjords.

Right:
A Sopwith 1½-Strutter, a spotter-reconnaissance aircraft, being launched from *Australia*. The section of the flying-off platform which fitted over the gun barrels was portable and could be removed. *IWM*

Below:
***Valiant* with flying-off platforms mounted on both B and X turrets. Her standard outfit of aircraft between the wars was one Sopwith Camel fighter and one Sopwith 1½-Strutter spotter-reconnaissance, the latter being a two-seat aircraft. Here the Camel appears to be on B turret and the 1½-Strutter on X. The B turret platform was removed in 1932 and that on X in 1937.** *MPL*

Right:
The Hawker Osprey, which replaced the Fairey IIIF and which was in service in the 1930s. *Fleet Air Arm Museum*

Below:
Fairey IIIF Mk IIIB. These aircraft were catapult launched. *Real Photos*

Bottom:
***Resolution* with a catapult fitted on X turret. The catapult was fitted in 1936 and could be extended prior to launching the aircraft, in this case a Walrus.** *IWM*

Left:
Rodney **was fitted with a catapult on X turret in 1936. Here a Walrus is being loaded.** *IWM*

Below:
A Walrus being launched from the athwartships catapult on ***Howe*****. The two hangar doors are visible on the right. One aircraft could be stowed in each hangar with another two on the catapult. The small turntable in the rails can be seen bottom left.** *IWM*

Bottom:
A Swordfish being recovered on to ***Malaya*****.** *IWM*

4
Principal Actions and Losses

Date	Action	Ships taking part	Losses
06 Sep 1870			*Captain*, capsized in storm
01 Sep 1875			*Vanguard*, sank after collision
11 Jul 1882	Bombardment of Alexandria	*Alexandria*, *Inflexible*, *Invincible*, *Monarch*, *Sultan*, *Superb*, *Temeraire*	
22 Jan 1893			*Victoria*, sank after collison
30 May 1906			*Montagu*, total loss after grounding on Lundy Island
28 Aug 1914	Heligoland Bight	*Invincible*, *Lion*, *New Zealand*, *Princess Royal*, *Queen Mary*	
27 Oct 1914			*Audacious*, mined
24 Nov 1914			*Bulwark*, ammunition explosion
08 Dec 1914	Falkland Islands	*Inflexible*, *Invincible*, *(Canopus)*	
01 Jan 1915			*Formidable*, torpedoed
24 Jan 1915	Dogger Bank	*Indomitable*, *Lion*, *New Zealand*, *Princess Royal*, *Tiger*	
18 Mar 1915	Dardanelles — first break-through attempt	Many ships took part in Dardanelles campaign at various stages, including: *Agamemnon*, *Albion*, *Canopus*, *Goliath*, *Implacable*, *Irresistible*, *London*, *Lord Nelson*, *Majestic*, *Ocean*, *Prince George*, *Prince of Wales*, *Queen*, *Swiftsure*, *Triumph*, *Venerable*, *Vengeance*	*Irresistible*, mined, *Ocean*, mined

Date	Action	Ships taking part	Losses
13 May-27 May 1915	Dardanelles — continuing operations		*Goliath*, torpedoed (13 May), *Triumph*, torpedoed (25 May), *Majestic*, torpedoed (27 May)
06 Jan 1916			*King Edward VII*, mined
27 Apr 1916			*Russell*, mined
31 May 1916	Jutland	Grand Fleet and 2 BS *Agincourt, Ajax, Bellerophon, Benbow, Canada, Centurion, Collingwood, Colossus, Conqueror, Erin, Hercules, Iron Duke, King George V, Marlborough, Monarch, Neptune, Orion, Revenge, Royal Oak, St Vincent, Superb, Temeraire, Thunderer, Vanguard* 5 BS *Barham, Malaya, Valiant, Warspite* Battlecruisers *Indefatigable, Indomitable, Inflexible, Invincible, Lion, New Zealand, Princess Royal, Queen Mary, Tiger*	*Indefatigable, Invincible, Queen Mary*, by gunfire
09 Jan 1917			*Cornwallis*, torpedoed
09 Jul 1917			*Vanguard*, internal explosion
17 Nov 1917	Off Heligoland	*Courageous, Glorious, Repulse*	
09 Nov 1918			*Britannia*, torpedoed
14 Oct 1939			*Royal Oak*, torpedoed
13 Apr 1940	Second Narvik	*Warspite*	
03 Jul 1940	Oran	*Hood, Resolution, Valiant*	
09 Jul 1940	Calabria	*Malaya, Royal Sovereign, Warspite*	

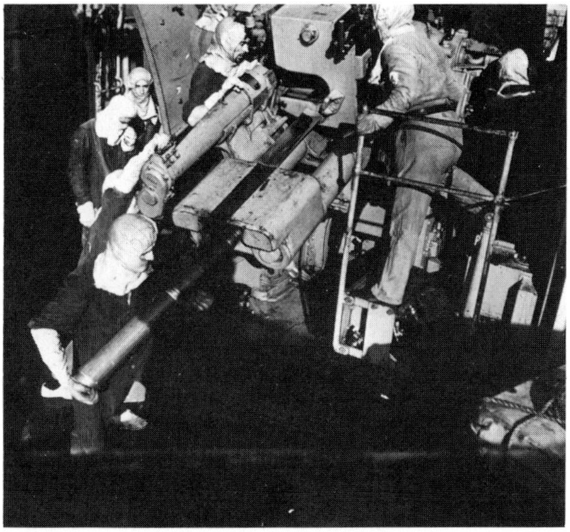

Left:
***Nelson* in action, 16in guns firing.** *IWM*

Right:
A 4.7in gun crew in action on board a 'Nelson' class ship.
IWM

Above:
***Malaya.** IWM*

Date	Action	Ships taking part	Losses
23-25 Sep 1940	Dakar	*Barham, Resolution*	
27 Nov 1940	Cape Teulada	*Ramillies, Renown*	
09 Feb 1941	Bombardment of Genoa/ Leghorn	*Malaya, Renown*	
28 Mar 1941	Matapan	*Malaya, Valiant, Warspite*	
21 Apr 1941	Bombardment of Tripoli	*Barham, Malaya, Warspite*	
18-27 May 1941	Hunt for *Bismarck*	*Hood, King George V, Prince of Wales, Ramillies, Renown, Repulse*	*Hood* (19 May), by gunfire
20-23 May 1941	Crete	*Barham, Queen Elizabeth, Warspite*	
24 Nov 1941			*Barham*, torpedoed
10 Dec 1941			*Prince of Wales* and *Repulse*, sunk by Japanese aircraft
08 Nov 1942	'Torch' — North African landings	*Duke of York, Nelson, Renown, Rodney*	
10 Jul 1943	'Husky' — Sicily landings	*Howe, King George V, Nelson, Rodney, Valiant, Warspite*	
09 Sep 1943	Salerno landings	*Nelson, Rodney, Valiant, Warspite*	
26 Dec 1943	North Cape	*Duke of York*	
06 Jun 1944	'Overlord' — Normandy landings	*Nelson, Ramillies, Rodney, Warspite*	
15 Aug 1944	'Dragoon' — South of France landings	*Ramillies*	
01 Apr 1945	Okinawa landings	*Howe, King George V*	
29 Jul 1945	Bombardment of Tokyo	*King George V*	

Bibliography

The True Glory, Warren Tute, Macdonald 1983.
The Battleship Era, P. Padfield, Granada 1972.
Rule Britannia, P. Padfield, Routledge & Kegan Paul 1981.
The Blindfold Game, E. Oakeshotl, Pergammon Press 1969.
Battleships, Anthony J. Watts, Macdonald & Jane's 1978.
British Battleships 1892-1957, Randolph Pears, Putnam 1957.
Battleship Design & Development 1905-1945, Norman Friedman, Conway 1978.
Royal Navy Battleships, W. D. G. Blundell, Almark 1973.
British Battleships, O. Parkes, Seeley Service 1973 edition.
The Big Gun, P. Hodges, Conway 1981.
The Navy at War 1939-45, S. W. Roskill, Collins 1960.
British Battleships of World War II, A. Raven and J. Roberts, Arms & Armour Press 1976.
Guns at Sea, P. Padfield, Hugh Evelyn 1973.
The Black Battlefleet, G. A. Ballard, Society for Nautical Research 1980.
Conway's All the World's Fighting Ships (three volumes), Conway.
Battleships and Battlecruisers, Tony Gibbons, Salamander 1983.
Battleships of World War II, Garke, Dulin and Webb, US Naval Institute 1980.
Battleships 1856-1977, A. Preston, Phoebus Publishing 1977.
Ironclads in Action, H. W. Wilson, Sampson, Low, Marten & Co 1896.
Battleships of World War I, A. Preston, Arms & Armour Press 1972.
Aircraft of the Royal Navy, Paul Ellis, Jane's 1982.
Atlas of Naval Warfare, Helmut Pemsel, Arms & Armour Press 1977.

Below:
***Irresistible* in 1914.** *Real Photos*

Index of Ships' Names

Achilles 18
Africa 62
Agamemnon (1883) 39
Agamemnon (1908) 63
Agincourt (1867) 22
Agincourt (1914) 81
Ajax (1883) 39
Ajax (1913) 78
Albemarle 59
Albion 55
Alexandra 30
Anson (1889) 44
Anson (1942) 98
Audacious (1870) 27
Audacious (1913) 78
Australia 73

Barfleur 51
Barham 84
Bellerophon (1865) 21
Bellerophon (1909) 70
Benbow (1888) 44
Benbow (1914) 79
Black Prince 14
Britannia 62
Bulwark 58

Caesar 53
Caledonia 18
Camperdown 44
Canada 82
Canopus 55
Captain 24
Centurion (1894) 51
Centurion (1913) 78
Collingwood (1887) 43
Collingwood (1910) 70
Colossus (1886) 42
Colossus (1911) 74
Commonwealth 62
Conqueror (1886) 40
Conqueror (1912) 74
Cornwallis 59
Courageous 93

Defence 16
Devastation 29
Dominion 62
Dreadnought (1879) 34
Dreadnought (1906) 66
Duke of York 98
Duncan 59

Edinburgh 42
Emperor of India 79
Empress of India 48
Erin 81
Exmouth 59

Formidable 57
Furious 93

Glorious 93
Glory 55
Goliath 55

Hannibal 53
Hector 17
Hercules (1868) 23
Hercules (1911) 74
Hero 40
Hibernia 62
Hindustan 62
Hood (1893) 50
Hood (1920) 94
Howe (1889) 44
Howe (1942) 98

Illustrious 53
Imperieuse 40
Implacable 57
Indefatigable 73
Indomitable 68
Inflexible (1881) 37
Inflexible (1908) 68
Invincible (1870) 27
Invincible (1908) 68
Iron Duke (1871) 27
Iron Duke (1914) 79
Irresistible 57

Jupiter 53

King Edward VII 62
King George V (1912) 78
King George V (1940) 98

Lion 77
London 58
Lord Nelson 63

Magnificent 53
Majestic 53
Malaya 84
Marlborough 79
Mars 53
Minotaur 22
Monarch (1869) 24
Monarch (1912) 74
Montagu 59

Nelson (1881) 33
Nelson (1927) 96
Neptune (1881) 36
Neptune (1911) 72
New Zealand (1905) 62
New Zealand (1912) 73
Nile 46
Northampton 33
Northumberland 23

Ocean (1866) 18
Ocean (1900) 55
Orion 74

Prince Consort 18
Prince George 53
Prince of Wales (1904) 58
Prince of Wales (1941) 98
Princess Royal 77

Queen 58
Queen Elizabeth 84
Queen Mary 77

Ramillies (1893) 48
Ramillies (1917) 88
Renown (1895) 52
Renown (1916) 91
Repulse (1870) 27
Repulse (1894) 48
Repulse (1916) 91
Resistance 16
Resolution (1893) 48
Resolution (1916) 88
Revenge (1894) 48
Revenge (1916) 88
Rodney (1888) 44
Rodney (1927) 96
Royal Alfred 21
Royal Oak (1862) 16
Royal Oak (1894) 48
Royal Oak (1916) 88
Royal Sovereign (1892) 48
Royal Sovereign (1916) 88
Russell 59

St Vincent 70
Sans Pareil 45
Shannon 32
Sultan 28
Superb (1880) 35
Superb (1909) 70
Swiftsure (1872) 28
Swiftsure (1904) 60

Temeraire (1877) 30
Temeraire (1909) 70
Thunderer (1877) 29
Thunderer (1912) 74
Tiger 82
Trafalgar 46
Triumph (1873) 28
Triumph (1904) 60

Valiant (1868) 17
Valiant (1916) 84
Vanguard (1870) 27
Vanguard (1910) 70
Vanguard (1946) 100
Venerable 58
Vengeance 55
Victoria 45
Victorious 53

Warrior 14
Warspite (1888) 40
Warspite (1915) 84

Zealous 21